WILDERNESS

Wilderness provides a multidisciplinary introduction into the diverse ways in which we make sense of wilderness: how we conceptualize it, experience it, interact with it, and imagine it. Drawing upon key theorists, philosophers, and researchers who have contributed important knowledge to the topic, this title argues for a relational- and process-based notion of the term and understands it as a keystone for the examination of issues from conservation to more-than-human relations.

The text is organized around themed chapters discussing the concept of wilderness and its place in the social imagination, wilderness regulation and management, access, travel and tourism, representation in media and arts, and the use of wilderness for education, exploration, play, and therapy, as well as its parcelling out in parks, reserves, or remote "wastelands." The book maps out the historical transformation of the idea of wilderness, highlighting its intersections with notions of nature and wildness and teasing out the implications of these links for theoretical debate. It offers boxes that showcase important recent case studies ranging from the development of adventure travel and ecotourism to the practice of trekking to the changing role of technology use in the wild. Summaries of key points, further readings, discussion questions, and Internet-based resources, such as short videos, allow readers to grasp the importance of wilderness to wider social, cultural, political, economic, historical, and everyday processes.

Wilderness is designed for courses and modules on the subject at both postgraduate and undergraduate levels. The book will also assist professional geographers, sociologists, anthropologists, and environmental and cultural studies scholars to engage with recent and important literature on this elusive concept.

Phillip Vannini is Canada Research Chair in Innovative Learning and Public Ethnography and a Professor in the School of Communication & Culture at Royal Roads University in Victoria, BC, Canada.

April Vannini received her PhD from the European Graduate School in Media and Communication and teaches at Royal Roads University in Victoria, BC, Canada, in the School of Communication & Culture as an Associate Faculty member.

Key Ideas in Geography

SERIES EDITORS: SARAH HOLLOWAY, LOUGHBOROUGH UNIVERSITY AND GILL VALENTINE, SHEFFIELD UNIVERSITY

The *Key Ideas in Geography* series will provide strong, original, and accessible texts on important spatial concepts for academics and students working in the fields of geography, sociology, and anthropology, as well as the interdisciplinary fields of urban and rural studies, development and cultural studies. Each text will locate a key idea within its traditions of thought, provide grounds for understanding its various usages and meanings, and offer critical discussion of the contribution of relevant authors and thinkers.

Published:

Nature
NOEL CASTREE

City
PHIL HUBBARD

Home
ALISON BLUNT AND ROBYN DOWLING

Landscape
JOHN WYLIE

Mobility
PETER ADEY

Migration
MICHAEL SAMERS

Scale
ANDREW HEROD

Rural
MICHAEL WOODS

Citizenship
RICHARD YARWOOD

Wilderness
PHILLIP VANNINI AND APRIL VANNINI

Forthcoming:

Space
PETER MERRIMAN

WILDERNESS

Phillip Vannini and April Vannini

LONDON AND NEW YORK

First published 2016
by Routledge
2 Park Square, Milton Park, Abingdon, Oxon OX14 4RN

and by Routledge
711 Third Avenue, New York, NY 10017

Routledge is an imprint of the Taylor & Francis Group, an informa business

British Library Cataloguing in Publication Data
A catalogue record for this book is available from the British Library

Library of Congress Cataloging in Publication Data
Names: Vannini, Phillip, author. | Vannini, April, author.
Title: Wilderness/Phillip Vannini and April Vannini.
Description: New York, NY: Routledge, 2016. | Series: Key ideas in geography | Includes bibliographical references and index.
Identifiers: LCCN 2015042493 | ISBN 9781138830981 (hardback: alk. paper) | ISBN 9781138830998 (pbk.: alk. paper) | ISBN 9781315736846 (ebook)
Subjects: LCSH: Philosophy of nature. | Wilderness areas–Philosophy. | Human ecology–Philosophy.
Classification: LCC BD581.V28 2016 | DDC 304.2–dc23LC record available at http://lccn.loc.gov/2015042493

ISBN: 978-1-138-83098-1 (hbk)
ISBN: 978-1-138-83099-8 (pbk)
ISBN: 978-1-315-73684-6 (ebk)

Typeset in Joanna MT
by Sunrise Setting Ltd, Brixham, UK

CONTENTS

List of illustrations

TABLES

BOXES

1

INTRODUCTION

Brady's Beach, Bamfield (Photo: April Vannini)

The most difficult aspect of writing a book is the first paragraph. Hours, days went by as new opening paragraphs were drafted, evaluated, and then quickly discarded. The writer's block seemed to drag on forever. Frustrated, tired, and anxious at the impasse we eventually resolved to put books, journals, and computers away and headed out for a leisurely walk in the woods hoping for inspiration.

It was an overcast autumn afternoon. The late September rains had begun wetting the forest floor after a long summer drought and the first wild mushrooms of the season, our favorite chanterelles, had sprung to life. Upon sighting their long-awaited arrival I (Phillip) ran back to our parked vehicle and grabbed a cloth bag in order to collect them. These delectable wild things can grow in unpredictable quantities wherever they please, without a care in the world for paths, trails, or any other instruments for channeling human mobility, so soon enough we were traipsing deep in the bush and filling our bag one mushroom at a time, careful enough to respect the forest's thick undergrowth with our steps, but determined to find and bring home as many as possible.

Two hours later, our reason for being in the bushes in the first place—curing writer's block—had dissipated in the silent darkness of the Pacific temperate rainforest. Our concerns with this book's introduction and with a million other mundane scholarly preoccupations had been rendered utterly meaningless by our primal hunger—well, appetite, really—for freshly harvested food and for a deeper, however momentary, connection with the wild. Our habitual tendencies to rush through the day's list of multiple tasks and chores had been slowed down, almost magically, by the slow-growing ancient cedars and hemlocks we walked around. It was in this way that, almost magically, our reason for bothering to write this book in the first place had resurfaced. Simply put: we love wilderness and all things wild. Therefore it was precisely in the wild—we realized upon our return to the office—that this book should set off.

But were we actually in the wild? Were we in a place we could properly call wilderness? Together with some 4,500 souls we call home a 56 km^2 island in British Columbia's Georgia Strait—one of the continentally renowned Canadian Gulf Islands, just north of America's equally famed San Juan Islands. The precise area we roamed in our search for mushrooms is a patch of undeveloped land, roughly 300 acres in size, owned by the Crown and currently subject to a long and somewhat controversial claim dispute with a local First Nation.

Well, "undeveloped," to a degree. There are many different paths and trails in the bushes, there are remnants of not-so-old human presence, as well as all kinds of traces of frequent human use: from a dozen geocaches scattered around to an abandoned washing machine in the middle of nowhere—a littering curiosity even marked by a dedicated trail. This particular Crown land, to be clear, enjoys no discernible environmental protection of any kind, yet its uncertain legal status makes it impossible for any one person or group to own, exploit, or develop. Walking around in it is not so clearly legal itself, as there are multiple signs at the edges of the land urging people to keep out (nearby there is also a sign informing us about the presence of a "Wilderness Watch" in effect) and even a metal bar preventing access—strangely standing right next to an inviting, official-looking trailhead.

In addition, how wild was our activity? Though we abandoned the trail, our car was never more than an hour's walk away. To make our foray into the bushes easier we had brought along with us more than a few trappings of civilization, beside our cloth bag: from sturdy hiking boots and rain-repellent clothing to a Swiss Army knife and even a GPS. Throughout the time we walked deep in the forest we could always hear the distant roar of automobiles driving on the island's main road, and the odd jet making its final approach into Vancouver's International Airport. Had we owned and carried a mobile phone with us, we might have even been able to send a text or post pictures of our fresh mushrooms on Facebook, maybe with a link to a recipe. As for the fungi, sure enough they seemed wild, but their presence amidst the moss and trees had only been possible through the choices local authorities had made to refrain from authorizing residential building—for the time being at least—and therefore to unwittingly protect the habitat which they needed in order to grow.

So, did this book truly begin in the wilderness? Or did it begin in a place and time only wild in an illusionary sense of the word—a place more knotted through with deeply binding social ties than with wild evergreen roots? To be honest we highly doubt a conclusive answer to those questions can be found in our book, or anyone else's—or at least one that will convince everyone. And yet, as the next pages will hopefully reveal, the very act of asking those types of questions—and of considering multiple and perhaps equally satisfactory answers—is something of immense value for our understanding of wilderness and wildness. But let us back up for a moment and begin by examining not so much what wilderness and wildness are—more on that shortly—but rather where they might be, and who might own, access, and control them.

THE NATURE OF WILDERNESS

In the beginning, everything was wilderness—we might be tempted to quip. But such a pronouncement would be rife with problems. The notion of wilderness as an incipient and pristine land free of human interference—with the implicit connotation that the onset of human civilization somberly marks nature's fall from grace—is so incredibly problematic and useless that it is undoubtedly a lot easier to begin from the present condition and then gingerly work our way backwards in time. In doing so we could say that today, in the most basic terms, the worldwide status of wilderness is that of a twofold entity. Wilderness is alternatively (1) an area protected by public authorities (or, arguably, private parties), or (2) an area that is not officially designated as such or legally protected, but is nonetheless considered to have *ad hoc*[1] wild characteristics. As the latter designation is a controversial one, let us describe officially protected wilderness first through some examples, and then show the necessity of expanding the first definition in order to add the idea carried by the second one.

In Canada, where the two of us live, wilderness is an officially recognized legal term. Canadian law recognizes wilderness through the definition and management of protected areas, which, following the International Union for Conservation of Nature (IUCN) (see Box 1.1), are understood as: "a clearly defined geographical space, recognized, dedicated and managed, through legal or other effective means, to achieve the long-term conservation of nature with associated ecosystem services and cultural values" (International Union for Conservation of Nature, 2008). As a result of this definition and under the aegis of Canada's Wildlife Act, at the present time the Canadian government holds massive expanses of allegedly "intact natural areas" that are renowned worldwide for their biological diversity and aesthetic appeal.

Box 1.1 THE INTERNATIONAL UNION FOR CONSERVATION OF NATURE

The International Union for Conservation of Nature (IUCN) was founded in 1948 as the world's first global environmental organization, and today it is the largest professional global conservation network. It encompasses 1,200 member organizations including over 200 governments and nearly 1,000 non-government organizations

(NGOs). Almost 11,000 voluntary scientists and experts based in 160 countries provide their services to the IUCN. They are supported by over 1,000 staff in 45 offices and hundreds of partners in public, NGO, and private sectors around the world. The Union's headquarters are located in Switzerland.

The IUCN describes itself as a "neutral forum for governments, NGOs, scientists, business and local communities to find practical solutions to conservation and development challenges" (International Union for Conservation of Nature, 2008). Its work is focused on thousands of field projects and activities around the world. It is funded by governments, bilateral and multilateral agencies, foundations, member organizations, and corporations, and it enjoys an "Official Observer Status" at the United Nations General Assembly.

The IUCN has created a very influential system for the classification of protected areas. The categories have been recognized by the United Nations and many countries' governments, and have therefore been incorporated into various forms of legislation. They are as follows:

1a: Strict nature reserve

Category 1a designates strictly protected areas set aside to protect biodiversity and also possibly geological/geomorphic features, where human visitation, use, and impacts are strictly controlled and limited to ensure protection of the conservation values. Such protected areas can serve as indispensable reference areas for scientific research and monitoring.

1b: Wilderness area

Category 1b protected areas are usually large unmodified or slightly modified areas—retaining their natural character and influence without permanent or significant human habitation—which are protected and managed so as to preserve their natural condition.

Continued

2: National park

Category 2 protected areas are large natural or near-natural areas set aside to protect large-scale ecological processes—along with the complement of species and ecosystems characteristic of the area—which also provide a foundation for environmentally and culturally compatible spiritual, scientific, educational, recreational, and visitor opportunities.

3: Natural monument or feature

Category 3 protected areas are set aside to protect a specific natural monument, which can be a landform, sea mount, submarine cavern, geological feature such as a cave, or even a living feature such as an ancient grove. They are generally quite small protected areas and often have high visitor value.

4: Habitat/species management area

Category 4 protected areas aim to protect particular species or habitats and management reflects this priority. Many Category 4 protected areas will need regular, active interventions to address the requirements of particular species or to maintain habitats, but this is not a requirement of the category.

5: Protected landscape/seascape

A protected area where the interaction of people and nature over time has produced an area of distinct character with significant ecological, biological, cultural, and scenic value; and where safeguarding the integrity of this interaction is vital to protecting and sustaining the area and its associated nature conservation and other values.

6: Protected area with sustainable use of natural resources

Category 6 protected areas conserve ecosystems and habitats together with associated cultural values and traditional natural resource management systems. They are generally large, with most of the area in a natural condition and a proportion under sustainable natural resource management, where low-level non-industrial use of natural resources compatible with nature conservation is seen as one of the main aims of the area.

Source: International Union for Conservation of Nature (IUCN) (2008). Protected areas. www.iucn.org/about/work/programmes/gpap_home/gpap_quality/gpap_pacategories (last accessed October 24, 2014).

Through a variety of federal departments (e.g. Environment Canada, Fisheries and Oceans, and Parks Canada) the Canadian government plays an essential role in the protection of such wilderness, chiefly by defining certain habitats as sufficiently important in terms of national ecological value. The designation of an area as "protected" consequently results in policies and programs that aim for the conservation of species at risk and migratory birds. To date, the network of protected areas totals 12.4 million hectares, 85% of which are classified as "wilderness areas."

Many countries around the world have a similar approach to defining and protecting wilderness. In the United States it is Congress that holds the power to designate an area as wilderness, following the legislation contained in the 1964 Wilderness Act (see Box 1.2). There are currently around 43.5 million hectares of land protected in America, about 5% of the land mass of the 50 states (a little over half of that land is found in Alaska alone). These 756 wilderness areas range in size from the five acres of Florida's Pelican Island to the nine million acres of Alaska's Wrangell-St. Elias wilderness reserve. As is the case in Canada and elsewhere, in the US wilderness protection results in public access restrictions and in strict regulations prohibiting mechanized mobility, development, and even many forms of recreation.

In Australia the network of protected areas—which may not always be synonymous with IUCN officially designated "wilderness" Category 1b areas—is known as the National Reserve System. A protected area is defined in the Australian context as "a clearly defined geographical space, recognised, dedicated and managed through legal or other effective means to achieve the long-term conservation of nature with associated

Box 1.2 THE US WILDERNESS ACT

The US Wilderness Act is one of the most historically meaningful pieces of legislation on the matter of wilderness worldwide. The document was chiefly written by Howard Zahniser from the Wilderness Society, though it was the cumulative effort of over half a century of environmental activism and the final iteration of over sixty drafts. It was signed into law by President Lyndon Johnson on September 3, 1964 and enacted by the 88th Congress.

The enactment of the Wilderness Act led to the formation of the National Wilderness Preservation System and to the protection of 37,000 km^2 of land. Today's 443,000 km^2 of protected wilderness are administered by the National Park Service, the US Forest Service, the US Fish and Wildlife Service, and the Bureau of Land Management.

The 1964 Act states that "a wilderness, in contrast with those areas where man and his own works dominate the landscape, is hereby recognized as an area where the earth and its community of life are untrammeled by man, where man himself is a visitor who does not remain" (US Wilderness Act, 1964, Sec. 2C). The US Wilderness Act (Sec. 2C) further seeks to capture the "character" of wilderness as a place that:

1. generally appears to have been affected primarily by the forces of nature, with the imprint of man's work substantially un-noticeable;
2. has outstanding opportunities for solitude or a primitive and unconfined type of recreation;
3. has at least five thousand acres of land or is of sufficient size as to make practicable its preservation and use in an impaired condition;
4. may also contain ecological, geological, or other features of scientific, educational, scenic, or historical value.

In sum, wilderness areas are supposedly large, socially valuable, "untrammeled," "natural," and "undeveloped" places (Landres *et al.*, 2005). These are qualities that have been the subject of much criticism, controversy, and reinterpretation.

Arguably the most emblematic of the controversies proved to be about the nature of roads. As Sutter (2009) has argued, the foundation of American environmental consciousness and the formation of national parks in the US has a great deal to do with the construction of roads into previously inaccessible natural areas and the opportunities for motorized recreation and related activities (e.g. camping) this infrastructure afforded to millions of Americans. Yet, the Wilderness Act defined wilderness areas as "roadless," at once restricting some types of access (e.g. automobile-based) and seemingly encouraging others (bicycle-based). Roadlessness, however, was found to be a vague criterion and was later taken to mean the absence of roads improved and maintained by mechanical means. This later interpretation led to the 1986 ban of bicycles from wilderness areas.

The Wilderness Act specifies broad qualities of the "character" of wilderness, but stops short of articulating them in precise, manageable detail. The concept of wilderness character is poorly understood, cuts across many resource areas, and has never been properly described or monitored, argued a team of managers and ecologists from several US Departments and Agencies in 2005 (Landres *et al.*, 2005; also see Carver, Tricker, & Landres, 2013). Their jointly written report, "Monitoring Selected Conditions Related to Wilderness Character" (Landres *et al.*, 2005), is a powerfully illustrative national framework that clearly operationalizes how wilderness qualities can be ascertained and enforced through day-to-day decisions.

ecosystem services and cultural values."[2] All in all, 10,000 protected areas amount to 127 million hectares. These lands are owned and managed by the Commonwealth as well as by states, territories, indigenous groups, non-profit conservation organizations, and even private farmers. Once designated a "protected area," stringent legal safeguards guarantee the perpetual conservation of a wilderness region. But in order to be classified as such the land must meet scientifically determined criteria that specify its environmental and cultural value.

The list could go on, but we will turn our attention to more international examples later. The point we wish to make for now—as the examples of Canada, the USA, and Australia show—is that in today's world wilderness is first and foremost a piece of land defined as a wilderness reserve which is protected by law and informed by scientific prerogatives and political priorities. But how much sense would it make to limit our definition of wilderness to such publicly owned, officially sanctioned reserves alone? Take the example of the Alladale Wilderness Reserve found in the Caledonian forest of northeast Scotland. Purchased by English millionaire Paul Lister in 2003 with the intent to manage it in order to "rewild" it, the Alladale Wilderness could not quite be counted as an official wilderness area because it is not publicly owned and managed. But even if we agreed that wilderness areas can be privately controlled, another conundrum would immediately appear: would the land have been considered a wilderness area prior to its purchase and designation as a wilderness reserve in 2003? If our answer is negative then we are faced with the paradox of a land becoming an official wilderness only after having been named and regulated as such, which is tantamount to making wilderness a human fabrication through and through. If our answer is positive then we must wonder why and how it is being rewilded. At what point exactly did it lose its wildness? And what made it less, or more, of a wilderness? And if it is currently being rewilded, has it been rewilded enough by now to be considered a wilderness? (For more on this see Marris, 2013.)

All of these questions—paradoxical-sounding but very legitimate indeed—compel us to understand wilderness not only as an officially recognized and protected area, but also as a place that features constantly changing qualities of wildness. It is in this sense that we must expand the working definition proposed earlier in order to make conceptual space for wilderness areas that are not officially designated as such or legally protected. Wilderness, in this latter sense, might very well be *a place that is perceived to be wild* by a person or group. Wilderness is therefore a highly diverse kind of place—ranging from something like a nondescript bush to a desert, or from the Arctic tundra to an overgrown backyard—that is *considered, claimed, believed, felt, imagined, or otherwise held* to be wild. This *ad hoc* type of wilderness is something that will complicate our treatment of wilderness immensely in this book, but at the same time it is also something that will make our subject matter all the more fascinating. What wilderness is across space and time, therefore, and more importantly how wilderness is achieved as an intersubjective and more-than-human accomplishment, will become one of our primary concerns. Wilderness, as it has already become apparent, will become entangled throughout this book with multiple human and non-human actors and closely related concepts, values, and practices that will shape its identity.

Chief amongst these concepts is that of nature. Wilderness is pristine nature, according to the dictionary. But nature is one of the most complex ideas in the English dictionary, as cultural theorist Raymond Williams (1984) has famously remarked. The idea of "pristine nature" is especially contentious because by this expression we typically refer to nature that is devoid of the presence of humans (Plumwood, 1998). Humans, apparently, are outside of nature and their interferences end up spoiling it. But how did human become a species that is outside of nature? Are we not animals too? Are we somehow above nature? Do we not ever catch a cold? Doesn't nature ever call us to the loo? Understanding wilderness in the broader context of our relation to nature (see Berry, 2010) is something that will keep us very busy in the chapters to come. Our hope is that by problematizing our shared idea of wilderness we will be able to show as many perspectives as possible on the conceptualization, experience, practice, and representation of wilderness.

IN SEARCH OF WILDERNESS

What has already begun to transpire is that wilderness is a highly contentious term (see Box 1.2; also see Callicott and Nelson, 1998; Nelson and Callicott, 2008). It should be no surprise, then, that over the past three decades, as its study has mushroomed around the world across varied academic fields, the idea of wilderness has stirred intense theoretical and political debate in the academy and outside, pitting those who believe that it stands as an ideal form of essential nature untouched by humans against critics who argue that the construction of the meanings of wilderness is informed by strong social forces that reveal important cultural and political dynamics.

We could try and begin to make sense of this debate by turning to the common dictionary. Wilderness, the dictionary informs us, originates in twelfth-century English from the juxtaposition of *wil*—meaning wild—and *deor*—meaning deer and more generally a beast or animal

(Nash, 2001). A different definition, however, can be found through an alternative etymology. According to Jay Hanford Vest (1985) the root of the word "wilderness" is "will" and "willed," as in will-force and willpower. This is something that does not escape Nash's (2001) reading, but Vest argues that the will in question is not the will of beasts but the will of the land. "Ness," in fact, is derived from Old Norse, Swedish, Danish, and Low German and refers to a promontory headland or cape. Wilderness may then simply denote a place that has a will of its own. See Box 1.3 for more on naming wilderness.

Etymologies aside, the most common contemporary denotations of wilderness are the following:

1. A wild and uncultivated region, as in a forest or desert, uninhabited or inhabited only by wild animals; a tract of wasteland.
2. A tract of land officially designated and protected as wilderness.
3. Any desolate tract, as of open sea.
4. A part of a garden set apart for plants growing with unchecked luxuriance.
5. A bewildering mass or collection.

Box 1.3 WILDERNESS: WHAT'S IN A NAME?

The word "wilderness" has a considerable amount of "baggage." More properly speaking, wilderness has several negative connotations for some people, and it is for this reason that a handful of critical scholars have argued for its abandonment. Nash (2001, pp. 1–3) offers a great deal of insight into the issue.

"Wild"—the root of the word "wilderness"—conveys the idea of being lost, confused, and disordered. Indeed the related adjective "bewildered" is synonymous with this condition. But why is wilderness related to these feelings? Who, in other words, would feel this way upon entering a place, such as a forest, where wild animals roam freely? An insider with intimate local knowledge of the land? Or more likely an outsider, such as an urban dweller, accustomed to viewing wilderness as Other? As Nash (2001, p. 2) puts it: "The idea of a habitat of wild beasts implied the absence of men [*sic*], and the wilderness was conceived as a region where a person was likely to get into a disordered, confused, or 'wild' condition. . . . The image is that of a man in an alien environment where the civilization that normally orders and controls his life is absent."

Given wilderness's historical baggage, Callicott (2008) argues that we would be better off replacing the term "wilderness" with the more scientifically precise "biodiversity." For Callicott the word "wilderness" connotes indelible traces of ethnocentric, androcentric, unphilosophical, unfashionable, genocidal, and unscientific ideas. At the core of these connotations stand powerful negative images, effects, and symbols (summarized in Table 1.1).

In contrast, the expression "biodiversity" is more neutral and alerts us to the value of preserving habitats for species that find them suitable for their survival. A focus on biodiversity would also lead to a clearer emphasis on the goals of a science of "reserve selection,

Table 1.1 The negative connotations of wilderness

Negative connotation	**Key symbol/symptom**
Ethnocentric	English colonist treating aboriginal homeland as useless and uncivilized wilderness
Genocidal	Forced removal of indigenous peoples from lands newly designated as parks
Androcentric	Lone male testing fortitude, virility, and other alleged masculine virtues outdoors
Racist	Wilderness as a frontier to be conquered as a proof of (white) national character
Unscientific	Historical ignorance about ecological relations between humans and environments
Unphilosophical	Christian Puritan view of wilderness areas as sites of sin and deprivation

Continued

design, and management" (Callicott, 2008, p. 372) and to the development of better criteria for the classification of biodiversity reserves.

However, abandoning the word "wilderness" is the wrong solution, according to others. Val Plumwood (1998), for example, wishes to preserve the term wilderness because of its potential to remind us of important historical lessons. Plumwood argues that wilderness has been traditionally defined as "virgin" land to be conquered, mastered, and controlled. These narratives reveal a great deal about androcentric attitudes toward nature and about the Western tendency to separate culture and nature—and thus to view the latter as the absence of the former. But instead of continuing to view wilderness as "empty," Plumwood argues, we simply need to reconceptualize it as a site for the *presence* of the Other: "the presence of long-evolving biotic communities and animal species which reside there, the presence of ancient biospheric forces and of the unique combination of them which has shaped that particular, unique place" (1998, p. 682).

Dictionary definitions are useful, but they can never be more than simple starting points. For a more descriptive definition we could turn to the precise and influential definition of the US Wilderness Act of 1964 (see Box 1.2). But would this truly suffice? Notions of wilderness areas as pristine places, such as that mandated by the US Wilderness Act, are popular with many environmental conservationists who, by casting wilderness as massive tracts of untrammeled land that are experiencing threats of gradual extinction, are generally able to convince lawmakers worldwide to warrant the protections afforded by wilderness preservation status (Lines, 2006; Milton, 1999; Proctor, 1998). And similarly they are also popular with those wishing to underline the growing commercial rarity of pristine wilderness areas worldwide, such as adventure travel operators and outfitters who wish to bank on the value of compellingly attractive, primitive places (see Swarbrooke *et al.*, 2003). But are these definitions responsible?

Not at all, we believe. All this talk of wilderness as a pristine state of nature free of regular human presence is nothing but a powerful discursive construction that obfuscates how nature is subject to contesting social forces, cultural studies scholars decry (e.g. Braun & Castree, 1998; Castree, 2005; Haraway, 1991; Latour, 1993; Macnaghten & Urry, 1998; Plumwood, 1998; Soper, 1995). One need not be a dyed-in-the-wool social constructionist to believe that the entire world's surface, even its most spectacular parks and preserves (e.g. see Miles, 2009), has been touched in one way or another by humans and that wilderness ranges on a continuum (see Carver, 1998). Legal environmental protection itself renders wilderness a human product—a profound irony that puts humans in the role of wilderness creators. In the words of environmental historian William Cronon, whose edited book *Uncommon Ground* (1996a) stands as the apotheosis of the constructivist position on this matter: "far from being the one place on earth that stands apart from humanity, [wilderness] is quite profoundly a human creation—indeed the creation of very particular human cultures at very particular moments in human history . . . . Wilderness hides its unnaturalness behind a mask that is all the more beguiling because it seems so natural" (1996b, p. 69; also see Cronon, 2008). Cronon (1996b, pp. 80–81) continues:

> The trouble with wilderness is that it quietly expresses and reproduces the very values its devotees seek to reject . . . . This, then, is the central paradox: wilderness embodies a dualistic vision in which the human is entirely outside the natural. If we allow ourselves to believe that nature, to be true, must also be wild, then our very presence in nature represents its fall . . . . To the extent that we celebrate wilderness as the measure with which we judge civilization, we reproduce the dualism that sets humanity and nature at opposite poles.

The trouble with the view of wilderness as pristine nature—which we will refer to as the "received view" of wilderness—thickens if we take a broader historical perspective. In their respective classic works *The Idea of Wilderness* and *Wilderness and the American Mind*, philosopher Max Oelschlaeger (1991) and historian Roderick Nash (2001) show that wilderness is a complex and ever-changing entity that has undergone extensive change over time and across space. As we will examine in greater depth in Chapter Two, wilderness is also a relatively recent idea: an outcome of worldwide population growth and residential and industrial development. When extensive tracts of forested land were widely available, wilderness was largely seen in part as menace and in part as yet-unexploited resource for expansion, or simply as useless wasteland. But as more and more Europeans and, later, North Americans began to lament the urban condition and the domination of humans over nature, wilderness grew in the popular imagination as a place of refuge and escape, a ground for moral and physical fortitude, and a site for the manifestation of the sublime (Nash, 2001; Oelschlaeger, 1991). The popularity of famed wilderness writers like Leopold, Muir, Thoreau, and many others serves as evidence of this historical shift in the attitude toward wilderness.

A great deal of recent social scientific literature has recently contributed to the further unsettling of the idea of wilderness as a pristine, "natural" environment. For example, Braun's (1997) research on the conflict between environmentalists and the resource-extraction industry in British Columbia's Clayoquot Sound has unveiled how wilderness emerges as a "discrete and separate object of aesthetic reflection, scientific inquiry, and economic and political calculation at particular sites and specific historic moments" (1997, p. 5). By denying the presence of culture in the wild, Braun (2002) shows that the received view of wilderness becomes a perilous colonizing force. Similarly, Suchet (2002) has observed in the Australian context how the legal designation of an area as wilderness has often resulted in the displacement of former human dwellers, especially aboriginal people (also see Hobbs, 2011).

Such a situation, which has led some observers to adopt the concept of "conservation refugees" (see Dowie, 2006; also see Spence, 1999), has taken particularly dramatic tones in Africa where the imperial visions of European colonizers first, and the conservation pressure of some international environmental NGOs (non-governmental organizations) later, has resulted in the formation of national parks and nature reserves that have deeply reshaped African ways of life (e.g. Neumann, 1995; Nustad, 2011). Neumann's historical analysis of the creation of Serengeti National Park, for instance, illustrates how outsiders' images of Africa as a wild continent reveal "a process of nature production rather than nature preservation" (1995, pp. 149–150).

The sustained analysis of practices of leisure in the wilderness (such as camping, see Kropp, 2009) can further contribute to our understanding of how the received view of wilderness can result in nature becoming a domain separate and distinct from society and everyday life. Such separation often results in the fetishizing of wilderness as a domain of experimentation with alternative roles, identities, and ways of being (Kropp, 2009)—thus reinforcing the idea that nature is "the Other" of normal social existence. And the status of wilderness as a fetishized and idealized site for escape is not lost on travel promoters worldwide. Nature-based tourism promotion, especially of destinations marketed as ultimate wilderness, thus frequently results in commodifying the wild and consequently pushing the quest for the next "secret" wilderness farther afield (Sæþórsdóttir, Hall, & Saarinen, 2011). As more of the world's "great outdoors" undergoes this cycle of gradual discovery, initial development, and more aggressive conquest, another way in which our paradoxical relation with the wilderness manifests itself is revealed. As long as wilderness remains apparently free of human presence its might and resilience seem compelling and omnipotent. But as soon as humans begin showing up, wilderness and nature suddenly transpire as weak and fragile, thus demanding that humans' presence be excluded while their distant protection be granted (see Hobbs, 2011).

More cultural studies of wilderness show that deconstructing wilderness as a domain separate from culture and society can result in the exposure of anthropocentric (Greig & Whillains, 1998; Plumwood, 1998), ethnocentric (Hodgins, 1998), and even androcentric ideologies (Sandilands, 2005). For example, Baldwin (2009, 2010) reveals how "wilderness narratives g[i]ve primacy to whiteness as a defining trait of national identity" (2009, p. 886) and Mawani (2007) outlines how park formation and management can work as an exclusionary exercise of juridical power over indigenous populations claiming rights over park land. The idea of wilderness has been so deeply embattled that arguably it might even be time to abandon the concept of wilderness altogether and move on.

Yet academic skepticism toward the received view of wilderness is not necessarily shared by the general public. As we will see in Chapter Three recent successful travelogues (e.g. Bill Bryson's *A Walk in the Woods*; Cheryl Strayed's *Wild*), films (e.g. Sean Penn's *Into the Wild*—an adaptation of the book by Jon Krakauer of the same name), documentaries (e.g. Werner Herzog's *Grizzly Man*) and countless TV shows following the exploits into the bushes by gold seekers, trekkers, climbers, and survivalists continuously add to the ever-growing pile of coffee table books, paintings, calendars, travel magazines, and skill manuals in forming a panoply of popular images and discourses. Wilderness may very well be a contested site within cultural studies, but it is still very much—and arguably more than ever—a prime site of popular culture performance (Kahn & Hasbach, 2013).

Furthermore, as outlined in Chapters Four, Five, and Six, the contemporary popularity and meaningfulness of such popular conceptions of wilderness worldwide has not escaped researchers all over the globe who study wilderness-based practices as diverse as "lifestyle sports" (e.g. Laviolette, 2007); youth camps (Van Slyck, 2006); logging, hunting, and fishing (Brandth & Haugen, 2005; Dizard, 1994, 2003; Lovelock, 2008a; Sandlos, 2008); art-centered hikes (Morris, 2011); trekking and mountaineering (Lund, 2005, 2008); bushwalking (Palmer, 2004b); kayaking (Waitt & Cook, 2007); motor-based mobilities (Jacques & Ostergren, 2006; Louter, 2009; Sutter, 2009); nature-based tourism (Olafsdottir, 2011, 2013a); traditional knowledge-based resource harvesting (Hunn *et al.*, 2003); the commodification of adventure (Cloke & Perkins, 2002); the domestication of wilderness landscapes for mass tourism promotion (Flad, 2009); wilderness therapy (Bettman *et al.*, 2013; Kelly, 2011; Marlowe *et al.*, 2002); the production and consumption of national parks (Baker, 2002; Carr, 2007, 2009; Catton, 1997; Hermer, 2002; Kopas, 2008; Neumann, 1999); and the multiple and contested meanings of mountains (Stoddart, 2012), glaciers (Cruikshank, 2006), forests (Kohn, 2013; Kosek, 2006; Marsh, 2009; Satterfield, 2003), and swamps (Ogden, 2001)—to name only the most visible practices.

It seems we are dealing with a double conundrum. Getting rid of the idea of wilderness would mean turning a blind eye to reality and ignoring wilderness's continued appeal as both a culturally meaningful idea and a "base datum"—a benchmark of sorts—for the approximation of ecological wholeness. But continuing to follow the received view would mean committing to perpetuating ignorance and injustice. What we need, we firmly believe, is a solution that will allow us to study wilderness as represented, practiced, experienced, and contested. Simultaneously we need to avoid taking normative sides on what wilderness is or is not, and what it should or should not be. We wish to be students of wilderness, in other words, not judges of its alleged authenticity.

The solution we propose is to be agnostic. Rather than a prescriptive definition such as the one implicit in the dictionary or in the US Wilderness Act we take an emic perspective on the matter and define wilderness as a *place that is considered to be a wilderness and treated as such*. That's it—let those who inhabit a wilderness be the judges! If someone, somewhere, somehow considers a place to be a wilderness our task should not be that of determining whether they are right or wrong, and why so. Rather, our job will be to examine how they have come to form such an orientation and how their orientation is manifested in practice and experience. We, in other words, will find wilderness wherever it is putatively found by others. Our agnostic definition of wilderness is inspired by the need to reconcile (rather than keep separate) academic cultural critiques on the *idea* of wilderness with the lived experience and the practices unfolding in wilderness *sites*. Therefore, to reiterate, throughout this book we will ask not only *what* is wilderness, but more precisely *how* is wilderness done. And relatedly: *where* is wilderness done (e.g. see Box 1.4)? In what ways is it sought after, engaged with, and experienced, and with what consequences—for both wilderness and wilderness seekers?

Box 1.4 DOES SIZE MATTER?

How large should a wilderness area be? Large enough to allow for a "two-weeks' pack trip" without encountering roads and cottages, as Aldo Leopold famously stated? Or, in less anthropocentric terms, large enough to allow for unconstrained large predator mobilities? And, conversely, how small can a wilderness area afford to be? Do small areas prevent most large predator species from surviving? And do small "islands" of wilderness surrounded by the rest of the populated world unduly allow for the development of residential and commercial areas right outside their borders? In sum, is there a minimum size for a wilderness area to be "truly wild"? To put all of these ideas in context let us look at the ten largest officially protected areas in Table 1.2. These are not necessarily IUCN Category 1b areas, but rather places recognized by many as having many qualities of wildness.

The sizes of the world's ten largest protected areas are undoubtedly impressive. But are hundreds of thousands of square kilometers necessary for wilderness to be authentically wild? It depends. Some conservation goals, such as the preservation of some rare

Continued

Table 1.2 The world's ten largest protected areas

Area	Country	Size in km²
Pacific Remote Islands Marine National Monument	US	1,271,500
Northeast Greenland National Park	Denmark	927,000
Chagos Marine Protected Area	UK (Indian Ocean)	640,000
Ahaggar National Park	Algeria	450,000
Phoenix Islands Protected Area	Kiribati	408,250
Papahānaumokuākea Marine National Monument	US	360,000
Great Barrier Reef Marine Park	Australia	345,400
Kavango–Zambezi Transfrontier Conservation Area	Southern Africa	287,132
Galapagos Marine Reserve	Ecuador	133,000
Great Limpopo Transfrontier Park	Southern Africa	99,800

plant populations, can be accomplished through small reserves. Other goals, such as the protection of resilient animal species not easily disturbed by human activities, are perfectly suited to average-sized multiple-use protected areas. In contrast, big carnivore protection requires massive tracts of protected land. Nonetheless, the "bigness" of wilderness exerts a unique fascination and appeals to both conservationists and the general public.

Influential conservation biologist Reed Noss (1998, p. 528) argues:

> [T]he desirability of large reserves, "bigness," is one of the few generally accepted principles of conservation. Protected examples of ecosystem types must be large enough to maintain viable populations of all native species and to persist in concert with natural disturbances. Large reserves are easier to defend against encroachment from outside, suffer less intensive edge or boundary effects, and require less management per unit area.

But how large should large be? Thomas (1990) argues that a population of 1,000 is adequate for the survival of species of normal population variability, but figures as high as 10,000 are necessary for long-term persistence of more highly variable animals. Translating these figures into land sizes depends on species behavior and types of habitat, and thus in turn on density and dispersion of the population in question. Noss (1998) believes that many of the world's reserves are still too small to serve their protection goals.

Yet, we must take into account the political and economic difficulties that the designation of large tracts of land for protection may cause (such as the possible removal of human residents or restrictions imposed on their traditional practices). And obviously we must account for the sheer size of land available to countries, as well. It is much easier for large countries and regions with relatively low population density (like Australia, the US, Canada, Brazil, Greenland) to set aside land, but not as easy for smaller, more densely populated ones. On this matter, transnational cooperation on protected area designation (such as the "Southern African" examples reported in Table 1.2) must be welcomed and encouraged.

THE VALUES OF WILDERNESS: A MULTIDISCIPLINARY AND GLOBAL APPROACH

Our intent in writing this book is not so much to list best practices for wilderness management, or take an ideological position on the ethics of wilderness, but rather to articulate in great depth how a particular kind of a place (a wilderness) and its qualities (of wildness) come to be—with all that this may imply for conservation, policy, cultural and historical awareness, and social and political relations. To achieve this broad purpose, we have written this book with a twofold strategy that encompasses a comprehensive review of the literature and a careful development of our original conceptual argument on wilderness as assemblage (see next section and Chapter Seven). This twofold strategy aims to prove useful to an audience of advanced undergraduate students in geography and related social sciences including sociology, cultural studies, environmental studies, and outdoor education. Of course we hope that our book will also appeal to many graduate students and scholars seeking a meaningful introduction to the topic. The extensive body of cited research and the analytical and conceptual observations we have developed throughout the book will no doubt prove invaluable to all these audiences.

Most books on wilderness are strongly anchored in national contexts. And because most writings on the topic are published by Americans, most volumes on wilderness are in one way or another about the US and about American ideas, histories, values, and perspectives of wilderness. In contrast, our approach is global. Our understanding of wilderness is comparative and hopefully multicultural, and our coverage of case studies in this text very diverse and international. Through this multi-faceted approach to the idea of wilderness and to its experience, practice, and representation we hope not only to appeal to a global readership, but also to be able to connect the question of wilderness with multiple and different social, cultural, environmental, political, and historical concerns ranging from climate change and overpopulation to exurbanization and the commodification of travel and leisure in the "great outdoors." Our *geographies*—in the plural—*of wildernesses* must begin, however, with a common approach and starting point. And that starting point, we believe, is our *care* and *concern* for the present and future of wilderness.

But why care about wilderness? It is quite common in the literature to embrace an apologetic approach about this matter. Even though the *idea* of wilderness has been rightly subject to multiple critiques—apologetic writers argue—the *places* where wilderness can be found are worthy of our tender loving care. We do not dispute the convenience of this approach, but we find it quite lame. A clean separation of any object, such as a person or a place, from its representations, such as ideas and beliefs about it, is impossible. It seems too much like telling someone we care about: "I love you for who you are, but I find the idea of you repulsive." That will not get us far. A much more intellectually healthy and constructive approach might be to accept the idea of wilderness and recognize its troubled past as much as its good moments. After all, even intimate partners have their nasty arguments and their bad times. The key is simply to live and learn to fail better next time—as Samuel Beckett might say.

But again, why, precisely, should we care about wilderness? Well, simply put, because "in wildness is the preservation of the world" (Thoreau, 2007 [1862], p. 26). What do we and Thoreau mean by this? The answer might be found in a comprehensive inventory of 30 arguments for the preservation of wilderness assembled by wilderness writer Michael Nelson (1998). Not everyone will buy each of the 30 arguments—backpackers might like some, preservationists will dislike others, hunters will favor a few, whereas scientists and naturalists will be drawn to select ones—but the menu is diverse enough that every one of us, indeed every reader of this book, will be convinced that there is value in the wilderness for just about anyone in the world, and that we should care for future generations' right to have at least as much of it as we ourselves do. In no particular order, Nelson (pp. 156–193) writes that wilderness is:

1. A repository of precious natural resources.
2. A rich hunting and fishing ground.
3. A vast source of species for natural medicinal use.
4. A self-regulating service "industry" for humankind (e.g. by working as a carbon sink).
5. A human and non-human life-supporting ecosystem.
6. A place to seek physical health and practice various kinds of therapy.
7. An arena for athletic and recreational pursuits.
8. A place to seek and practice mental health.
9. An aesthetically rewarding environment.
10. A source of artistic inspiration.
11. A quasi-religious place for spiritual and mystical encounters with the divine and the transcendental.
12. A scientific field "laboratory" for the observation of natural processes.
13. A base datum, or comparative standard measure, for land health.
14. A reservoir of biodiversity.
15. An outdoors "classroom" for many different kinds of learning.
16. A memorial site commemorating human evolution.
17. A cultural landscape.
18. A symbol of national character and identity.
19. A place for self-realization.
20. A buffer zone or "disaster hedge" that protects humans from unknown viruses and bacteria.
21. A bastion of individual freedom.
22. An optimal location for the study of past myths and the generation of new myths.
23. A metaphysical necessity for our understanding of civilization.

24. A site for the practice of minority rights.
25. A liminal space for the formation of small group social bonds.
26. The prime site for animal welfare.
27. The prime site for the smooth functioning of the earth, or Gaia, as an organism.
28. An inheritance to be preserved for future generations.
29. A place full of enchantment, wonder, and unknowns.
30. A place that has its own intrinsic value, regardless of human interests.

No matter which one of these 30 reasons we might agree with, wilderness has clear extrinsic value to many people, animals, and plant species, as well as intrinsic value. The anthropology and cultural studies students amongst our readers might be more interested in some arguments, whereas the students of sociology, history, and political science might be persuaded by others. Environmental studies might focus on some of these points more than others whereas practitioners in outdoor education, recreation, or geography might very well be inclined to believe in different ones. As a result, our book will try and speak to the value of all these *ideas and places* for all dwellers—temporary and permanent—of wild spaces.

Now, we feel compelled to remind you that our task as reviewers of the literature is not as easy as it might seem. If we had chosen to limit our review only to studies relating to IUCN Category 1b—the most "official" global definition of wilderness—our book would have been very, very short. There simply are not many contemporary studies, and especially studies by geographers and other social scientists rather than natural scientists, on "official" wilderness. As a result our approach had perforce to be ecumenical—in other words, broad and comprehensive. This approach is also quite in line with our agnostic and putative definition of wilderness. But it also created a bit of a problem for us. Like we said, we view wildness as something that can potentially unfold in any number of different wild places, from protected areas to the bush just outside of town. And yet, do believe us, there are infinitely many more studies on protected areas, national parks, wild recreation areas, and more-or-less officially designated (though not necessarily in IUCN Category 1b terms) areas than there are on the nondescript bush at the edges of town. Such is a limitation of the field of studies, and not of our intent or abilities in reviewing.

WILDERNESS AS ASSEMBLAGE

As it has by now become obvious, wilderness is a heterogeneous, amorphous entity. If wilderness can be found in the most remote edges of the planet as much as in freshly rewilded rural and even urban spaces around the corner, and if the quality of wildness depends on the wild thing itself as much as on the mind, eyes, ears, mouth, nose, and hands and feet of the beholder, then our laissez-faire "agnostic" approach requires a hermeneutic strategy that should make our task of inventorying, contextualizing, and interpreting diverse wildernesses around the world a bit easier. The concept of "assemblage" shall then be the primary tool (see DeLanda, 2006). It is a concept that will guide us regularly throughout this book—though sometimes overtly and sometimes subtly—so it behooves us to introduce it here in clear, unequivocal detail.

Simply put, an assemblage is a group, an aggregate, an assembly, a collection of things and/or persons. Take the example of a constellation like the Big Dipper. The Big Dipper is an assemblage of seven of the brightest stars in the Ursa Major constellation. The stars are joined together by the act of drawing an imaginary set of lines that unite them to one another, resulting in the shape of a common dipper. In this way the seven stars—Alkaid, Mizar, Alioth, Megrez, Phecda, Merak, and Dubhe—acquire a common identity and a shared subjectivity that ends up preceding them and defining them (who amongst us, after all, knows their names by heart?).

The same happens with wilderness—for each wilderness area may be thought of as a constellation comprising places, inanimate objects, animals, humans, ideas, memories, regulations, technologies, and much more. Take the Galapagos, for example, arguably one of the planet's most renowned and most protected ecosystems—an environment that has profoundly shaped conservation science as much as the global social, historical, and cultural imagination. Who amongst us can promptly roll out the names of all the dozens of islands that comprise this archipelago, from Baltra and Isabela to Floreana and Daphne Major? Yet through a simple act such as the formation of the Galapagos National Park a definite wilderness—the assemblage of many different islands and disparate permanent dwellers, migratory species, visitors, landscapes, regulations, etc.—comes to life.

The point is the same for wildernesses as it is for celestial constellations: by joining things together we can create assemblages through powerful relations that give form to clear and powerful images and identities. But as useful and powerful as these images and identities are, they are also deceiving in that they obscure the heterogeneity of their parts and the evolving nature of their relations. Take the Galapagos Islands again. One thing is Santa Cruz Island with its 20,000 permanent human residents, for example, and another thing is human-uninhabited Genovesa Island, dwelled on "only" by families of birds and plants. The point we want to make is simple: wilderness is always, and inevitably, an assemblage—whether we are talking about Alaska's Arctic National Wildlife Refuge or the little Crown land on our island. Wilderness is the outcome of the practice of drawing lines, establishing relations, setting up boundaries, and naming a place. Our task as students of wilderness is then to try and make sense of the relations that bind a wilderness place together, and distant wilderness areas with one another.

To make things even clearer, let us formulate another example. The Scottish Highlands are famous worldwide for their beauty, but much less notorious, at least internationally, is the Scottish practice of Munro bagging. Video 2—which we will share in Chapter 4—will give more information on Munro bagging, but for now let us present Munro bagging as a type of assembling. Munros are Scottish mountain peaks, 3,000 foot high or more. "Bagging" is another word for collecting. Munro bagging, therefore, is a recreational activity that consists of climbing each one of the 282 Scottish Munros. Now, how is this a type of assembling? Consider this: Munros had first to be "invented." 3,000 feet is a very arbitrary number that someone, notably Sir Hugh Munro (1856–1919), had to consider somehow meaningful. Think of how arbitrary that threshold is by reflecting on its equivalent in meters: 914.4. Inventing the Munros therefore meant settling for a number that made the challenge serious but not impossible (think of how different the challenge would be if the cap had been set at 2,000 feet, or 4,000 feet, or, say, 997 meters). Furthermore, it also meant amassing them, or, in better words, naming them and listing them as tokens of the same type. This act of assembling resulted in the formation of an official "table" of Munros which had to be vetted by the Scottish Mountaineering Club.

Is this list "natural" and a simple "matter of fact"?—you might wonder. Hardly. For what makes a peak such? Have you ever stood on the very summit of a mountain and looked a few feet away, at a rock formation just a few meters shorter than the summit you are standing on, and yet thought of it as very much a "peak" of its own, perhaps even harder to climb than the one you conquered? Indeed defining the peak of a mountain requires formal definitions and classifications. And that requires negotiation and debate. Initial tables listed as many as 538 Munros, but later on a new expression was created by Scottish mountaineers, "tops," to designate mountain peaks that were not quite a topographically prominent mountain of their own but rather a mere secondary protuberance. Debates are not all there is to this, of course. Mountain altitudes have got to be validly and reliably measured, but the accuracy of those measurements depends greatly on precise tools and their correct use. So over time new Munros have been added and subtracted, with some "tops" promoted to Munros and some Munros relegated to mere "tops." We are sure you are starting to see how what may appear as a fixed nature at first sight is instead very much the outcome of practices, discourses, and changing relations. But the extent of the Munro assemblage goes further too.

Drive or walk around anywhere during spring and summer in the Scottish Highlands and you will be surprised—especially if you come from North America—by the sheer number of climbers and hikers out and about the many mountain trails. Not everyone is walking in order to bag, of course, but the very idea behind "the Munros" as a clearly recognizable recreational entity has obviously had a great impact in popularizing the practice of hiking around the Highlands. The changing technological times we live in have had an impact on the growing popularity of this pastime too, with more and more walkers taking photographs and videos of their outings and uploading them on the Internet, or perhaps ticking off Munros and tops from any of the websites that dedicate themselves to helping hikers stay on top of things (pardon the pun). Next thing you know, of course, the rising popularity of this arbitrary constellation of mountains translates into very real business for guides, outfitters, hoteliers, restaurateurs, car rental companies, etc. We are sure you start to see how the Munro assemblage is not just a list of topographically prominent mountains, but a meshwork of relations that entangle hikers, businesses, organizations, and the myriad rocks and pebbles that form well-defined trails up the mountains. Like the Munros and Munro bagging, wilderness is very much a set of relations arising from various assembling activities.

Given what we just said about wilderness as the outcome of an act of assemblage, are we then joining the ranks of social constructionists who advocate that humans are the creators of everything, including nature? Well, not quite. As much as we believe in social construction, we believe just as much in relational ontology: the idea that realities, of any kind, are established through relations. And no self-respecting relational ontologist would ever say that the only agents capable of establishing relations are humans. Relational ontologies go both ways: from the animate (e.g. humans) to the inanimate (e.g. material objects and ideas) and vice versa. Our view—what some might label a "flat ontology" (see e.g. Bogost, 2012; Bryant, 2011)—therefore does not put human beings at the center of anything. We create wilderness as much as wilderness makes us.

Within the social sciences, and geography in particular, views like ours have come to be known as "more than human." As Robbins and Marks (2009) summarize, from a more-than-human perspective assemblages are believed to shape humans as much as humans shape them. For example, they write:

> Gardens are understood to *constitute* the subject status of gardeners (Power, 2005) and not simply be constituted by them. Research on conservation argues that bears *participate in* and contribute to non-governmental organizational actions (Hobson, 2007). Elm trees and tree diseases are both understood to *labor* in the American urban economy, affecting and altering capital accumulation (Perkins, 2007). Prions interact with scientists and cattle to *produce* uncertainties that impinge on processes of policy-making (Hinchliffe, 2001). Dogs make people (Haraway, 2003). Mosquitoes speak (Mitchell, 2002). In a recent frenzy of insightful "critical animism," non-humans have re-entered the landscape, confounding "social" explanations of anything and everything. On the other hand, prions, elk, gardens, bears, elm trees, mosquitoes and dogs do not exist or "act" in the world independent of socialized knowledge, discourse, and scientific text. In each newspaper invocation, in every laboratory observation and inscription, and in all debates about the "natural" or "unnatural" character of technology, a host of deeply humanized and politicized representations occur. This represents a puzzle for any project that admits social geographies are more than human: confronting a world constituted by more-than-human actors, joined in a cat's cradle of physically grasping relationships, threaded through a fun house of representations.
>
> (p. 177)

Such is the puzzle awaiting our attention in the chapters to come: a confrontation of a more-than-human world of wildernesses in constant formation and in ever-changing material, embodied, forceful, conceptual, sometimes symmetrical and balanced, and other times asymmetrical and unbalanced, relations.

We will dedicate more time and space in the book's concluding chapter to the notion of assemblages—as well as, of course, throughout every chapter—but for now let it suffice to add that our relational ontological view aims to surpass the old and tired battle between realists and idealists, on-the-ground conservationists and deconstructing intellectuals which often results in people talking past each other and getting angry (see Callicott & Nelson, 1998; Nelson & Callicott, 2008). A more-than-human approach to wilderness dismisses neither the power of ideas and histories nor the innate force of fungi, trees, and torrents; thus it makes space for all objects and humans with their differing capacities and ways of life. From this point of view, borrowing from philosophers Gilles Deleuze and Félix Guattari (1987, p. 25), "there is no longer a tripartite division between a field of reality (the world) and the field of representation (the book) and the field of subjectivity (the author). Rather, an assemblage establishes connections between certain multiplicities drawn from each of these orders."

CHAPTER OVERVIEW

This book is divided into seven chapters. The six chapters following this introduction focus on the history and philosophy of wilderness, its representations in the media, its experience and practice, its contentious development and exploitations, its conservation policies and management, and finally its rethinking—or reassembling.

Chapter Two, "Thinking through wilderness," shows how wilderness has been a subject of the social imagination and the collective generation of ideas about nature and culture for a very long time. Chapter Two examines the role of wilderness in the humanities (such as writing—both fiction and creative non-fiction—and current philosophical thought) and explores how throughout history we have come to understand and conceptualize wilderness and the ideas surrounding it (including ideas of nature, rurality, wildness, and wild landscapes) in different ways. In that chapter we focus in particular on concepts pertinent to "natural" spaces and the wild. We also examine the work of North American historians such as William Cronon and Max Oelschlaeger and the philosophical underpinnings of deep ecology, as well as some of the newer contemporary thought pertinent to the subject. We argue that the idea of wilderness continues to interest students of the humanities as it constantly inspires the generation of new theoretical concepts and ways of life.

In Chapter Three, "Representing wilderness," we examine ways in which wilderness is and has been represented in discourses for and by the "general public." We examine various media such as film and television, marketing and advertising, popular literature, visual arts, and other forms of popular culture where wilderness is represented, displayed, and sold. We expose how through these discourses wilderness becomes a commodity and we reflect on ways in which it is framed as a site for adventure, discovery, and exploration. We survey previously published empirical research and cases that discuss key media discourses on wilderness and "the great outdoors," the commodification and spectacle of wilderness, and the performance of the wild. We also reflect on how these discourses have changed in light of increasing awareness of climate change and environmental sustainability.

Chapter Four, "Experiencing and practicing wilderness," explores the ways in which wilderness is experienced by its regular users, dwellers, and one-time, short-term visitors. Through a survey of empirical research we explain how wilderness is used in multiple and often competing ways as a vehicle for outdoor education, therapy, resocialization, recreational use, and survival camps. We reflect on the relationship between wilderness and the self and identity, highlighting its supposed transformational potential. We also tease out the various ways in which wilderness is experienced by different individuals and lifestyle groups, who each bring their own perspective and attitude to wild spaces. In that chapter we also examine the notion of living full-time in the wilderness as is typically the case for a few individuals and communities whose work responsibilities, alternative lifestyles, or long-established cultural traditions necessarily unfold in wilderness areas.

Chapter Five, "Conserving and managing wilderness," examines the environmental politics surrounding the science of conservation and preservation of wilderness spaces, with a special emphasis on the consequences of growing environmental awareness and climate change worldwide. In that chapter we explore how wilderness spaces are fought over in law and policy-making arenas, and how competing discourses, social movements, and socioeconomic forces clash over the definition of wilderness areas. We also examine the ethics surrounding wilderness, with a special focus on deep ecology, biophilia, current philosophies of sustainability, and ecofeminism. We survey how wilderness is managed and administered in various ways across the world, considering the cases of nature reserves, parks, managed and unmanaged forests, and other types of protected lands and marine zones. We consider how climate change understanding and related policy debates have impacted wilderness areas. We also extend our analysis to aboriginal political issues and discuss how the definition of wilderness areas has traditionally entailed land-use conflicts with indigenous users.

In Chapter Six, "Utilizing and exploiting wilderness," we concentrate on how many of the practices unfolding in the wilderness and various experiences of the wild can easily cross over into exploitations of wilderness areas for human pleasure and material gain. Chapter Six examines empirical case studies that detail how tourist flows, natural resource exploitation, travel interest groups, and business and government pressures for development and exploitation make use or attempt to make use of wilderness areas across the world. Thus we examine how venturing into and out of the wilderness may at times leave a large footprint on wilderness spaces, even while under the guise of preservation and conservation.

Finally, in Chapter Seven, "Reassembling wilderness," we draw together the various threads outlined in the previous chapters of the book, we make some conclusions about the state of wilderness in the beginning of the twenty-first century, and outline various areas for further inquiry. Rather than introducing new material this chapter summarizes key ideas and highlights key theoretical contributions to the study of wilderness, aiming to arrive at somewhat final considerations on what wilderness is and what its futures might be. In this chapter we also further investigate our idea of wilderness as assemblage, and as meshwork, highlighting the usefulness of these metaphors for multidisciplinary understanding of wilderness and for social, cultural, and political debate.

SUMMARY OF KEY POINTS

- Wilderness may be an illusion, but it is also real in many ways. The mark of a wilderness area resides in its quality of wildness. However, what that quality actually is remains open to debate and contestation.

- Wilderness is alternatively (1) an area protected by public authorities (or, arguably, private parties), or (2) an area that is not officially designated as such or legally protected, but is nonetheless considered to have *ad hoc* wild characteristics.
- The idea of wilderness has stirred intense theoretical and political debate in the academy and outside, pitting those who believe that it stands as an ideal form of essential nature untouched by humans, against critics who argue that the construction of the meanings of wilderness is informed by strong social forces that reveal important cultural and political dynamics.
- Believing that wilderness is pristine or untouched nature is highly problematic.
- Each wilderness area may be thought of as a constellation comprising places, inanimate objects, animals, humans, ideas, memories, regulations, technologies, and much more.

DISCUSSION QUESTIONS

1. When was the last time you were in the wilderness? Where was it? What was wild about it?
2. What is the nearest wilderness to you? What makes it a wilderness?
3. Can wilderness be privately owned? Are wilderness reserves better off in the hands of the public or private groups?
4. What laws are in place for setting aside wilderness reserves in your country? Are these laws effective?
5. What is the meaning of nature? How does it differ from the meaning of wilderness?
6. Why has the idea of wilderness been criticized? Do you find these critiques to be fair?
7. Should we look for a different word for wilderness? If so, what word?
8. Can you add any reasons additional to the list of 30 arguments about why we should care about the wilderness?
9. Can you make a list of all the various social and non-human actors that play a role in making a particular wilderness of your choice?

KEY READINGS

Nash, R. F. (2001). *Wilderness and the American mind*. Fourth edition. New Haven, CT: Yale University Press.

Oelschlaeger, M. (1991). *The idea of wilderness: From prehistory to the age of ecology*. New Haven, CT: Yale University Press.

WEBSITES

International Union for Conservation of Nature (IUCN): www.iucn.org
The International Journal of Wilderness: http://ijw.org/
The WILD Foundation: www.wild.org
The Wilderness Committee: www.wildernesscommittee.org
The World Wilderness Congress: www.wild.org/main/world-wilderness-congress
Wilderness.net: www.wilderness.net

All of these websites were last accessed on December 2, 2015.

NOTES

1 For our specific purposes we use the term *ad hoc* wilderness instead of the otherwise plausible term de facto wilderness since de facto wilderness is already established in the literature as: "Public lands that are wilderness in the general sense of the term, roadless and undeveloped, but which as wilderness have not been designated by Congress. Lands potentially available for wilderness classification" (Hendee & Dawson, 2002, p. 610).
2 For more information see www.environment.gov.au/topics/land/nrs/about-nrs/requirements (last accessed December 2, 2015).

2

THINKING THROUGH WILDERNESS

The way ahead, Scotland (Photo: April Vannini)

In June of 2014 the Chilean government officially rejected a plan to build five hydroelectric dams in rivers located in the south of the country. For several years up to that point residents of Patagonia and environmentalist groups from all over the world had been lobbying the Chilean government against the project which, if approved, would have given the green light to HidroAysen to cut through massive swaths of forest land and submerge 14,000 acres in order to dam the flow of the Baker and Pascua rivers. Many in Patagonia, in Santiago, and in the northern hemisphere rejoiced at the announcement. But how could this decision, and the resistance that fueled it, have happened? We do not need to look too far back in time to find strong worldwide political support for dam building and hydroelectricity development as a clean and renewable energy-generating strategy. So, how could collective perceptions of something believed to be valuable—such as renewable resource utilization (and wilderness preservation)—change so quickly and so dramatically? The material we discuss in this chapter will help us understand what ideas lie behind cases like this and therefore allow us to appreciate how precise historical geographies of wilderness have unfolded over the past two centuries.

More precisely, we ask, how have ideas of wilderness changed across modern times? How have historical events and social, political, economic, and cultural processes affected how different people in different regions and countries understand and value wilderness? And in what ways have literary and philosophical thought shaped the idea of wilderness in and out of the academy? Throughout this chapter we will address these and related questions by tracing the evolution (albeit, a far from linear one) of the notion of wilderness through a variety of currents of thought from the humanities, politics, environmental history, philosophy, and social and cultural theory. We will survey—and, alas, inevitably condense and simplify—the thought and actions of globally renowned figures and less-renowned, but equally insightful, leaders, thinkers, and writers.

We have a lot of ground to cover and in order to be coherent and systematic we have organized this chapter in a relatively linear historical way. Though we fully recognize that ours is going to be a simplification of a complex and nuanced history we have identified four key "moments" in the development of the idea of wilderness over the past two-and-a-half centuries. Each of these key moments, we argue, was either spurred by a series of foundational events, or embodied in the culmination of a particular process, or epitomized by an influential writing and publication. We are aware this is an artificial and elementary way of ordering history, yet it is a tidy and convenient one that will serve the purpose of illustrating a wide variety of material in an orderly and easy-to-understand fashion. The four moments—romanticization, legalization, contestation, and de/reconstruction—are presented in chronological sequence, though the beginning of each does not necessarily mark the demise of the previous one. Without further ado, we begin with wilderness the sublime.

WILDERNESS THE SUBLIME

The word "sublime" provides us with a perfect synthesis of the collective disposition toward wilderness typical of the modern era and especially of the Romantic movement of the late eighteenth and early nineteenth century. "Sublime" refers to something that impresses the mind and body with grandeur and power, something that is awe-inspiring, and worthy of being venerated and elevated in language and all forms of expression. Wilderness the sublime is wilderness that is subject to effusive praise, contemplation, and adoration. Wilderness the sublime is overpowering, enlightening, stimulating, and larger than life itself. Wilderness the sublime is the object of endless representations extolling its virtues, from poetry to landscape art, from philosophy to religious mysticism (Nash, 2001 [1982]).

In his monumental treatise on the history of the wilderness idea Max Oelschlaeger (1991) explains in precise detail how wilderness became sublime. Oelschlaeger (1991), like Nash (1982), finds that the sublimation of wilderness can be largely attributed to mutating geohistorical mechanisms of supply and demand. As European and American civilization continued to expand and settle in ever-increasing large cities and towns, the great expanses of forested land untouched or virtually untouched by humans which previous generations had taken for granted became increasingly rare (see S. Stoll, 2007). The creation of the first national park in the world, Yellowstone in 1872, works as evidence of this changing attitude in America. "Cities were built," writes Oelschlaeger (1991, p. 4),

> canals dredged, forests cut and burned, grasslands fenced, and the land brought into use for crops and cattle; native animals like the bison were systematically hunted nearly to extinction, partly to promote farming and ranching, partly to facilitate 'pacification' of the native children of the wilderness—the American 'Indian'.

But while some viewed wilderness as merely as an obstacle to the advancement of urbanization and development, or merely an economic resource to tap into, others began to think of the value of wilderness in its own right, as something to treasure and preserve. The spur to wilderness conservation fueled by this early process of sublimation and romanticization still continues strong today, and it is no accident that one of the foremost proponents of this view, Henry David Thoreau (1817–1862), continues to enjoy immense popularity in the present.

While Thoreau focused on nature and wilderness in a vast variety of publications, it is in *Walden* that most of his core ideas can be traced. *Walden* is the product of a unique experiment: a 26-month sojourn in a wood cabin situated by Walden Pond, a rural but relatively undeveloped location near Concord, Massachusetts. During his stay Thoreau worked in the field and on his cabin, hiked daily, and contemplated the passing of the seasons and natural processes as he mused about the relationship between civilization and wilderness. Famously, Thoreau observed that the hardships of self-reliance in the wilderness are not a burden, but rather an exercise in living simply and in appreciating the basic pleasures of life. In his words:

> I went to the woods because I wanted to live deliberately, to front only the essential facts of life, and see if I could not learn what it had to teach, and not, when I came to die, discover that I had not lived. I did not wish to live what was not life, living is so dear, nor did I wish to practise resignation, unless it was quite necessary. I wanted to live deep and suck out all the marrow of life, to live so sturdily and Spartan-like as to put to rout all that was not life, to cut a broad swath and shave close, to drive life into a corner, and reduce it to its lowest terms, and, if it proved to be mean, why then to get to the whole and genuine meanness of it, and publish its meanness to the world; or, if it were sublime, to know it by experience, and be able to give a true account of it in my next excursion.
>
> (Thoreau, 1962 [1854], pp. 172–173)

As one can glean from this passage the message of *Walden* is simple, and it is because of its simplicity that it continues to resonate today. By traveling to the wilderness one can gain precious reflexive distance from the shallowness and materiality of civilization, from the ways it spoils the spirit, dulls the mind, and weakens the body. Wilderness allows its seekers to rediscover life's essentials: its responsibilities and its modest but immensely rewarding pleasures of food and shelter. By stripping oneself of the customs, pretensions, and illusions of modern culture, one can find within wilderness truth, beauty, justice, and virtue (Dean, 2007).

Wilderness the sublime, for Thoreau and so many inspired by *Walden*, has an organic appeal capable of awakening us from the quiet desperation brought about by modern living. Wilderness the sublime is where true freedom can be sought: freedom from the grip of civilization on the ego, and freedom to be at one with the laws of the natural world and the timeless meanings contained therein. The quiet and solitude that can be practiced within the wilderness is instrumental for the achievement of clarity, knowledge, and resolve. Quiet and solitude allow one to find delight in the qualitative immediacy of the evening, of a pond, of a pine needle, of an encounter with a deer (Dean, 2007). Skeptical of the scientific quest to explain and thus contain the sublime power of wildness, in *Walden* Thoreau urges us to transcend "Judeo-Christian presuppositions about time, the scientific idea of nature, Cartesian dualism, and the Baconian dream: wilderness is neither an alien enemy to be conquered nor a resource to be exploited but 'the perennial source of life'" (Oelschlaeger, 1991, p. 158).

As a matter of fact, to truly understand the romanticization of wilderness, it is to the scientific mind that we must turn. As Oelschlaeger (1991, p. 68) argues, it was through "science, technology, and liberal democracy [that] modern people hoped to transform a base and worthless wilderness into industrialized, democratic civilization." The modernist scientific outlook on nature aimed to contain and explain everything formerly thought to be mysterious and magical. Following the Enlightenment, the modernist attitude toward reality indeed put a premium on reason and the scientific method, thus turning to wilderness as an object of dispassionate inquiry. Human mastery of nature was crucial not only to reinforcing an anthropocentric sense of security and dominance, but also to combating a competing secular worldview that conceived of nature as God's creation.

The modern scientific outlook on nature succeeded in propagating a mechanistic view of nature and confining aesthetic and poetic attitudes to the realm of folklore, the fine arts, and the humanities. However, in purging such attitudes from science modernism also made it possible for these attitudes to gain a powerful appeal as critical cultural and intellectual currents. In both literature and philosophy Romanticism thus flourished as a reactionary movement which valorized a personal, intimate, and emotional connection with the vitality of nature. Rather than conceptualizing nature as a machine as the scientists did, the Romantics therefore viewed it as an organism. Oelschlaeger (1991, p. 99) explains:

> To the Romantics nature was not a lifeless machine, mere matter-in-motion, but a living organism created by divine providence; they believed that God's presence was revealed through an aesthetic awareness of nature's beauty. To the Romantics the scientific idea of nature-as-matter-in-motion was sterile, objective, and stultifying. The poetic view of nature gravitated toward its wild and mysterious aspects, the felt qualitative rather than measured quantitative dimensions of experience, known through immediate contact rather than through experimentation. Feeling instead of thinking, and concrete emotion rather than abstract conception, were the essence of the Romantic awareness of nature.

Wilderness the sublime was thus born. While previous generations of Romantic writers and artists had turned to the serene and orderly beauty of city gardens to sing the praises of nature, eighteenth- and nineteenth-century Romantics learned to appreciate the chaotic and unrestrained power of the wild (Nash, 1982). Edmund Burke, for example, associated the sublime with fear of the wild, but instead of a debilitating fear he spoke of awe and admiration. In Europe, a change of attitude toward the Alps served as a perfect epitome of this transformation. Previously loathed and dreaded for their raw ugliness, a new generation of wilderness enthusiasts followed Jean-Jacques Rousseau in his newly found enthusiasm for the Alps' remoteness, strangeness, and mysteriousness. Like Thoreau did in the bushes of New England, it was in the solitude of Alpine and similar environments that European Romantics found an escape from over-civilized, boring, and stale society. It was in wilderness that they could cultivate a primitivist way of life bent on the cult of simplicity and the purity of the soul.

For some Romantics, not all, the soul referred to the spirit: the ghostly spirit intended in the traditional, religious sense of the word. Wilderness the sublime, for them, brought the spirit closer to God as the creator of the universe. This attitude seems easy to take for granted today, but it was radically new then (M. Stoll, 2007). Christianity had for a long time juxtaposed civilization against nature and posited that religious revelation was instrumental to ending humans' primitive wildness. Until deist orientations to nature became popular, many Christians viewed the wilderness as a dangerous and dark hell on earth where witches, uncivilized savages, and wild beasts roamed free and unpunished. Romantic deism, instead, looked to the wilderness as pure nature untouched by the corrupting hand of men. It was through wilderness that God showed his might. It was in wilderness that his vision of beauty and his idea of life could be observed (M. Stoll, 2007).

This reading of history—from wilderness the sublime and Romanticism as a manifestation of and reaction toward modernism, to deism, primitivism, and so on—makes a great deal of sense if we assume that the world is exclusively composed of European people and European migrants to America. But it is not. In fact, from an indigenous or aboriginal perspective, the very idea of wilderness made little sense at this historical junction (see Perreault, 2007; Chief Luther Standing Bear, 1998). Wilderness the sublime was born in modern European and North American cities, argues Nash (1982). It was "the literary gentleman wielding a pen, not the pioneer with his axe" (Nash, 1982, p. 44; also see S. Stoll, 2007), nor the aboriginal, who learned to view the wilderness as Other and thus as a negation of his urban way of life. The space that appeared as foreign, alien, and exotic to the urban literati was, however, a familiar place for the countless and nameless dwellers of communities that had learned to live not *in spite of* wildness, but *with* it (Perreault, 2007; Chief Luther Standing Bear, 1998). And indeed "it" was no wilderness. "It" was a river that would provide fish, a bush that could yield medicine, and a valley that could grow heat in the form of firewood. "It" was not a nameless swath of land awaiting discovery, development, protection, or conquest. "It" had names, names that had been passed on from generation to generation to share information about what went on there, what was available, and how daily life could take place there (Chief Luther Standing Bear, 1998). Such a worldview and way of life was at times ignored by eighteenth- and nineteenth-century European colonizers, and at times entirely misunderstood as a result of myths and stereotypes (see Box 2.1).

While the culturally specific idea of the sublime could not and should not be transplanted outside of the Western world, it is interesting to realize that in the East the relationship between pristine nature and humans was similarly marked by admiration and enchantment (Nash, 1982). Jainism, Buddhism, and Hinduism called for compassion toward all living things

Box 2.1 ROMANTICIZING AFRICA'S NATURE

Romantic ideals of wilderness and nature are not mere literary traces free of material consequences. Romantic ideals played an instrumental role in actually *producing* nature both at home and abroad. Neumann's (1995) research on the establishment of Serengeti National Park provides us with an in-depth look into the ways Romantic ideals and early British colonialism reshaped African landscapes and ways of life. Neumann's key argument—one that has informed much of the postcolonial literature since—is that the continental view of African nature as pristine and empty wilderness was a dangerous colonial invention which eventually resulted in the displacement of thousands of African people.

Drawing from Hobsbawm and Ranger's (1983) theory on the social construction of tradition, Neumann argues that British colonial agents invented African traditions for the African people. Neumann (1995, p. 151) explains:

> on the one hand there was a romanticization of pre-European African society which included ideas of moral innocence, a respect for African bush-skills, and a generalized notion of the noble savage, a mythological construction which Europeans evoked repeatedly in colonialist encounters with India, Africa, and the Americas. On the other hand there was the modernizing mission . . . whereby Africans would be freed from their backwardness and become efficient producers within the sphere of the British colonial economy.

These ideologies directly translated into policies and programs directed at changing the way local inhabitants interacted with the land, ranging from the provision of education directed at increasing agricultural productivity to wildlife conservation schemes aimed at curbing hunting. The creation of national parks throughout the continent, where wildlife could find a safe sanctuary from

Continued

the double threat of development and extermination, was a direct consequence of these policies.

Right as the parks were formed British preservationists argued that African people would be allowed to live within their boundaries as long as they practiced a "traditional" way of life. This was a unique concession. British Romantic notions of landscape—rendered popular by landscape paintings of artists such as Lorrain and Poussin—made a clear distinction between productive rural landscapes, where human labor was the chief focus, and consumptive landscapes which managed to adhere to an aesthetic of idyllic nature. "The introduction of these spatially distinct ideologies, consumption and production, preservation and development," writes Neumann (1995, p. 153), "had major ramifications for the transformation of African land rights and land-use practices. The process of development and preservation entailed nothing less than the recasting of the African landscape, a phenomenon inseparable from the recasting of African society."

In fear of losing Africa's "Eden," European preservationists set out to preserve wilderness through a system of parks meant to conserve "primordial, undisturbed, unchanging and pure" nature (Neumann, 1995, p. 154). In light of this attitude, tolerating people within these parks made sense because prevailing European ideas of African traditional culture cast African people as living in a natural state, not unlike a species of fauna. But this would turn out to be a double-edged sword, as those African residents whose ways did not fit the European myth of a primitive and traditional life could not be allowed to live within park boundaries, regardless of land rights. Large numbers of Maasai in particular were thus evicted from Serengeti National Park, unless they could somehow freeze social change and economic development within their communities in order to fit the colonial conception of a traditional life.

and humans were thought to be an inextricable element of nature. As in the deism of modern Europe and North America, Nash (1982) argues, Eastern thought venerated wild nature as the essence of God. Thus both Chinese and Japanese sought unity with the universe within natural landscapes, cultivated sacred rhythms, and practiced respect and love. The Shintoist attitude toward wilderness, in particular, worshiped mountains, torrents, forests, and storms (Nash, 1982)—still recurring elements in Japanese iconography to this day. Art was often the outcome of protracted sojourns in undeveloped settings where Chinese and Japanese painters, philosophers, and writers meditated, practiced quiet and stillness, and sought inner and outer harmony.

While it is impossible to do any sort of justice to the complexity of these philosophies in the limited space we have (for more on Eastern philosophies and nature see Callicott & McRae, 2014) it should prove interesting to report here on a story drawn from the early Buddhist tradition. It is a story shared to motivate followers of the Buddhist doctrine to seek freedom from suffering and to pursue enlightenment and liberation. It is a story about Siddhartha, a prince of the Gotama clan. Siddhartha was born to be a ruler and he was raised accordingly. He lived in luxury, learned the royal arts, was taught ability, strength, and influence from an early age, and was carefully insulated from the rest of the world to ensure he would not be polluted by it. Despite the fact that he had everything, he was not satisfied. He felt empty and unfulfilled. He found that suffering could not be avoided by accumulating wealth and enjoying earthly pleasure. But even this understanding did not come easy and Siddhartha had to struggle with pain and depression for some time as he continued to tackle possible solutions. "Then," McLeod (2014, pp. 92–93) continues the story:

> [O]ne day, Siddhartha had another experience that again helped him progress in his thinking and on his path. He saw a wandering mendicant monk begging for food, wearing ochre robes symbolizing the retreat from the worldly lives of most people. Such wanderers, Siddhartha had learned, studied, meditated, and searched for the truth about human life and the world, distant from everyday life in the city, wandering in the wilderness. Perhaps, Siddhartha thought, this is the way to find the answer to the problem of suffering. If there is an answer to be found—if it is possible to escape the human condition, it must be in the wilderness in contemplation and not in the worldly life that it is to be found. Involvement with pleasure and duties cannot be conducive to learning the answer and overcoming suffering. So one night Siddhartha left his wife and child and all of

> his possessions behind and fled into the dark night, going deep into the wilderness. . . . Siddhartha wandered alone for some time, practicing, contemplating, and meditating, looking for the answer to the problem of suffering. But nothing was forthcoming, and his frustration grew. Was there simply no answer after all? Then, one day, he had enough. He vowed to either find the answer or wither away and die. He decided to stay underneath the peepal tree he had taken shelter under the day before and not to move from there until he solved the problem of suffering once and for all. . . . After days of contemplation and meditation, almost miraculously, Siddhartha discovered the key to ending suffering, and became "enlightened." He finally understood that there was a path to end suffering, and that this path consisted in a number of steps of self-cultivation. This revelation came through gaining an understanding of what causes suffering, and the way we can undermine the creation of suffering. Siddhartha's own suffering slowly disappeared, and then underneath the peepal tree his face brightened, revealing the enigmatic smile we know today on the figure famous across the world, represented in millions of statues, paintings, and arts of all kinds. Siddhartha had become an enlightened one. He had become a Buddha.

MAKING A FEDERAL CASE OUT OF IT—WILDERNESS LEGALIZED

Neither Thoreau nor any of the Romantic writers, artists, and philosophers had a particularly practical or actionable vision for protecting wilderness. While their work inspired millions—and continues to do so posthumously—the fine arts and the belles-lettres they generated had only an *indirect* political effect by way of reshaping the public opinion and inspiring leaders to whom the task of ground-level action would eventually be left. Two of the most influential of these leaders were John Muir and Aldo Leopold, individuals whose contribution to the history of wilderness could not be overstated. Muir and Leopold not only published numerous and at times sophisticated essays that contributed to the development of the wilderness idea, but were also instrumental in the foundation of important environmental organizations (Muir the Sierra Club, Leopold the Wilderness Society), and contributed to the creation of laws and policies that directly led to the preservation of land and wildlife.

John Muir (1838–1914) is rightly considered the grandfather of the American conservation (later known as preservation) movement, and the father of the US national parks, for his contribution to the political struggle to set aside land for no fewer than six parks (the best known being Yosemite and Sequoia). Born in Scotland and transplanted into California, Muir deeply shaped how Americans viewed nature and wilderness through both his political activities and his numerous and accessible writings. Muir's view on wilderness was influenced by Emerson, Thoreau, transcendentalists, and Romantics, as well as by primitivist writers. But Muir did more than echo and popularize his predecessors. While he was no systematic philosopher, Muir was instrumental in critiquing the dominant anthropocentric view of nature so deeply popular at the time and positing, instead, a kind of early biocentrism founded around an organismic view of nature. Muir was also an avid trekker and scientific observer, as well as a religiously devout man whose deist views deeply influenced his Christian philosophy.

Similar to Muir, Aldo Leopold (1887–1948) combined literary and philosophical writings (his most popular book, *A Sand County Almanac* (Leopold, 1986), has sold more than two million copies to date), scientific work (in conservation ecology and forestry), and political activism. Leopold was also an academic, a University of Wisconsin professor in game management within the agricultural economics department. A bird enthusiast and avid hiker, Leopold played a leadership role in establishing the Gila Wilderness in New Mexico in 1924, the world's first officially recognized wilderness area. Leopold worked for years in the American Southwest for the US Forest Service, for which he developed—amongst other projects—a comprehensive management plan for the Grand Canyon. While, like Muir, Leopold was no systematic philosopher, his writings on what he called a "land ethic" and on forestry management have had a very long legacy. By a "land ethic" Leopold meant a biocentric system of thinking that blended ecological principles of species interrelatedness, a Romantic aesthetic sensitivity toward landscape, preservationist policies, and moral values. In 1921 Leopold also defined wilderness in a way that exerted a tremendous amount of influence over the development of the US Wilderness Act some four decades later. Wilderness, he wrote, should be understood as "a continuous stretch of country preserved in its natural state, open to lawful hunting and fishing, big enough to absorb a two weeks' pack trip, and kept devoid of roads, artificial trails, cottages, or other works of man" (cited in Nash, 1982, p. 186).

It is important to remain cool-headed about the impact of central figures like Thoreau, Muir, and Leopold (Robert Marshall could be added to the pool as well). Any student of the historiography of wilderness will promptly realize that the popularity of these men has increased exponentially over the years—so much so that nowadays no history of wilderness could be properly written without paying due homage to these leaders. Yet, we must be careful about mythologizing them. Thoreau, Muir, and Leopold have been as influential as they have because their activism and ideas struck a chord with the changing spirit of the times (we must keep in mind, however, that Thoreau died without an inkling of how popular one day he would become). So, without excessively lionizing particular individuals, let us examine what is so uniquely important about this "moment" in the history of wilderness—a moment loosely coinciding with the period between the birth of Yellowstone National Park and the passing of the US Wilderness Act.

Though it would be an exaggeration to state that wilderness is an American invention, it is true that Americans were the first to take political action to make wilderness a legal reality and that their actions later influenced many others around the world. The legalization of wilderness which occurred at the end of the nineteenth century, and then continued throughout the twentieth, stemmed from at least two factors. The first was a distinctive culture of exceptionalism. Americans at the time knew that their young nation lacked the "high culture" and traditions of Europe. It would have been difficult for any American to gain much pride and confidence in their national identity by playing against Europe at Europeans' game: their arts, their cities, their history, and their distinctive achievements. It was much easier for Americans, however, to find a sense of exceptional distinctiveness in their vast and largely underdeveloped landscape. American wilderness stood against Europe's gardens and manufactured rural landscapes as a symbol of a young nation thirsty to tackle challenges, capitalize on the courage of the pioneer spirit, and invent itself by pulling itself up from the bootstraps. As Nash (1982, p. 69) puts it:

> American nationalists began to understand that it was in the wildness of its nature that their country was unmatched. While other nations might have an occasional wild peak or patch of heath, there was no equivalent of a wild continent. And if, as many suspected, wilderness

> was the medium through which God spoke most clearly, then America had a distinct moral advantage over Europe, where centuries of civilization had deposited a layer of artificiality over His works.

The preservation of wilderness, therefore, was an essential step in the recognition of North American culture and heritage.

The second factor contributing to the legalization of wilderness—both in and outside of North America—was the changing dynamic in the supply and demand of undeveloped land, which we briefly mentioned earlier. Though undeveloped land was still abundant at the end of the nineteenth century, development pressures were starting to mount everywhere. Growing cities needed resources like water and electricity. Changing nations also demanded a wider transport infrastructure and the expansion of railroads, and later highway networks, had begun to threaten areas that had hitherto remained out of people's reach. In the US all of these forces began to collide in unprecedented strength over the fate of Hetch Hetchy Valley: a site as dear to the burgeoning environmentalist movement as it was to San Francisco city planners and developers (Nash, 1982). While the Hetch Hetchy battle was lost and a new dam was built—resulting in the flooding of a section of northwestern Yosemite National Park—newly organized environmental interests coalesced around the issue of wilderness protection in a way that would define American environmental policy for decades.

Simultaneously, tourism development had been growing steadily. In both America and elsewhere in the world, the influence of Romantic writings on nature had begun to manifest itself in an increasing appetite in people for outdoor recreation. Increasing members of the bourgeoisie had begun to explore the "great outdoors" in search of natural curiosities, panoramic landscapes, and fancy mountain resorts where they could see other people and be seen by them. All of this led US President Grant to give national park status to Yellowstone on March 1, 1872, out of the fear that the land's unique attractions—geysers, hot springs, waterfalls, etc.—would be acquired and exploited by private interests at the expense of the greatest number of people. Far from barring leisure seekers, however, the laws instituting the new national parks resulted in greatly facilitating tourist development. The parks had been protected not *from*, but *for* the people.

Fueled by the invention and popularization of the automobile, parks in America and around the world in fact became greater and greater sources of financial revenue for a blooming tourist industry. As the twentieth century wore on, gone were the times, especially in Europe, where outdoor recreation necessitated a "two weeks' pack trip." Lodges, huts, and refuges began to sprout in mountain areas in the Alps and later in the American Rockies, dotting landscapes already marked by booming resort towns. Europe's greater population density, of course, compounded the problem. With regard to the American case, Turner (2012, p. 24) observes:

> In the early twentieth century, when Henry Ford put the Model T into mass production, the American "road trip" was born. Each summer, tourists loaded up their new automobiles and drove them into the great outdoors, camping by roads, streams, and high mountain meadows, and they set their sights on national parks and forests. In the 1920s the Park Service and Forest Service began to cater to the auto-tourists in hopes of attracting more visitors and funds from Congress. Often, the two agencies competed with each other to promote their lands. The Forest Service, in particular, feared that new parks would be carved out of the national forests. Thus, the agencies built roads, laid out new campgrounds, and established mountain resorts.

The historical research of scholars like Sutter (2009), Turner (2012), and Miles (2009) is essential for our collective understanding of this period in global, and especially American, history. The legalization of wilderness, according to all three, arose out of the realization that leaving wilderness undefined and at the mercy of piecemeal development would lead to its eventual extinction (Kormos, 2008). The rise of global organizations like the Sierra Club, the Wilderness Society, the International Union for Conservation of Nature (IUCN), and similar groups worldwide, according to Turner (2012, p. 25), can then be understood as the urge to care for nature as much as a countercultural reaction to consumerism, capitalist development, and tasteless mass recreation. Nonetheless, the champions of wilderness maintained that protected land should exist first and foremost for the enjoyment of outdoorsmen. Whether by horse or by canoe, whether for fishing and hunting or trekking and camping, whether for spiritual retreat or aesthetic inspiration, wilderness protection ought to be administered to allow their use by recreation seekers. Miles (2009, p. 4) explains:

> The bottom line, though, was that parks were, as the Yellowstone legislation said, "pleasuring-ground for the enjoyment and benefit of the people." Initially an American gentry could, in relative comfort, visit protected remnants of the wild frontier and, for a time, live out their Wild West fantasies. A national park is an American institution, a product of the idealistic impulse in a largely individualistic society to offer something for the common good.

It would be impossible to understand how wilderness became the precise bureaucratic reality it is today without a close attention to this period. Because national parks were so clearly formed for the common use and enjoyment of their users, Miles (2009) argues, and because parks were developed so quickly with roads and lodges allowing automobile tourism, it became clear that the large areas of American national parks still unreached by development were a different entity altogether. Still very much part of nature, still very much part of the "great outdoors," the park lands that were still too difficult to reach by road, too remote to allow for any form of development, and too forbidding for the common tourist, were initially perceived as "wastelands" (Miles, 2009, p. 3): useless spaces that retained their original "wildness." As a result of the passing of the Wilderness Act, in the US the National Wilderness Preservation System (NWPS) was set up to maintain these lands free of permanent modifications, free of human control (other than the indirect control exercised by legislated protection), and free of human intervention. And other lands were soon added to the NWPS.

The definition of wilderness, in a sense, thus emerged as a negative reality: an opposition and a denial of all that is human. As we know very well today, no landscape is likely to ever have been free of human intervention after the arrival of humans on the planet, but at the time the belief in the existence of pristine spaces was both common and convenient—convenient especially because it allowed groups of activists to make a convincing case for the protection of places whose uniqueness could be easily lost and never regained. The consequence of this schism is that today many people across the world live in countries that have legislation protecting parks and wilderness areas and that the rules are quite different in many cases (Kormos, 2008). So, by the time the US Wilderness Act was passed in 1964 and by the time many other similar acts were written into law worldwide, wilderness had become a thoroughly legal reality with a precise definition that often went against the common usage of the word "wilderness." Legally, wilderness had to be a place utterly free of human presence and intervention, mostly free of development, and restricted in access. And the burden of proof was on any group keen on preserving such a land. While legalization brought clarity, the protection afforded by legal designation demanded the erasure of human existence in both the present and the past (otherwise the land could no longer count as "virgin") (for more see Chapter Five; also see Spence, 1999). And it also demanded that nature, not culture, reigned supreme.

WILDERNESS CONTESTED

The growing efforts to define, regulate, and manage wilderness which occurred throughout the last decades of the twentieth century had both positive and negative outcomes (Wuerthner, Crist, & Butler, 2014). The most obvious positive outcome is that vast tracts of land became officially recognized as important to international, state, and regional governments' conservation priorities and therefore protected for generations to come. But the protection afforded by legislative bodies worldwide had numerous shortcomings and negative consequences as well. As the twentieth century ended and a new millennium began, a growing number of critics began to coalesce around a nexus of different problematic issues. In this section we will examine three of these coalitions and their critiques: the anti-environmental movement, deep ecology, and critical race/"Third World" criticism.

In order to understand the anti-environmental movement we must first understand conservationism, upon whose principles many of the wilderness protection policies around the world were built (see Stradling, 2004; Wellock, 2007). Conservationism as a broad and diverse school of thought on environmental management is as old as the writings of John Muir and Gifford Pinchot (Wellock, 2007). Simply put, conservationism hinges on principles of sustainable resource use. The keywords here are "use" and "resource." When legislations around the world protect wilderness areas they often employ discourses that emphasize the importance of land management for the benefit of future generations of people (see Kormos, 2008). Benefits may be non-consumptive (e.g. recreation, nature appreciation) or consumptive (e.g. sustainable logging, hunting, fishing, mining, etc.), but regardless of their nature the rationale of such kinds of legislation remains the same: a land's resources ought to be managed for their instrumental value for the use and enjoyment of the people. The business of nature conservation and management nowadays is a very subtle one uniting government agencies, private service land contractors, non-governmental organizations, professional specialists, and even university faculties—and at times it is not so easy to distinguish ecocentric preservation from the more instrumental resource-oriented neo-liberal conservationist model, but the differences are there.

Conservationism is largely scientific, managerial, and administrative in nature, driven by broad scale costs-and-benefits analyses and bottom-line policies. To a great degree, conservationism is the inevitable outcome of state intervention in nature's affairs, or at least the inevitable outcome of the kind of utilitarian rationalization and bureaucratization that must answer to the logic of "investment" in natural resources (Wellock, 2007). Because conservationism is a bureaucratic and therefore political approach to wilderness protection, it is subject to shifting partisan agendas and priorities as well as to market-driven demands (at least in capitalist economies). It is therefore not uncommon for conservationist-inspired legislation to be subject to "exceptions" or legislative amendments which may at times render nature protection a smokescreen or a mere greenwash. "Conservationism" may therefore be a bit of a misnomer—superficially connoting more of a "land ethic" than it really does. "Resourcism," perhaps, might be a better-fitting name (though see Foreman, 2014, on the important differences between these two ideologies and movements). Resourcism more directly conveys the idea that nature has an immediate utility to humans by being amenable to technology-facilitated transformation and the satisfaction of changing needs and wants, and that ecological science and practice must therefore be directed at ensuring the sustainability of this exploitative relationship (Oelschlaeger, 1991).

Not to be confused with the resourcist ideology is the anti-environmentalist movement, the most visible manifestation of which is arguably the American "Wise Use" movement (see Echeverria & Eby, 1995). The Wise Use movement is a coalition of ultra-conservative and right-wing interest groups that became quite popular in the United States in the

1990s. The Wise Use movement lobbies for the dismantling of wilderness legislation and much environmental protection under the guise of free access to public land and justice for rural populations. In the US, Wise Use movement followers—which include extractive industry lobbies, logger organizations, groups of farmers and ranchers, property owners, off-road vehicle users, populist libertarians, and land development companies, amongst others—advocate for expanded use of public land by private and commercial parties and claim to be the most authentic of environmentalists in light of their everyday ties to the land. Similar attacks on resourcism and conservationism have been lobbied by other libertarian and right-wing coalitions around the world. This critical movement is interchangeably known as "anti-environmentalism" or "Green backlash" (see Beder, 2001).

On the other side of the political spectrum, conservationism is simply not sufficient, according to a cadre of environmental critics comprising followers of preservationist, biocentric, ecocentric, and deep ecology principles. Conservationism is too politically prudent, much too tied to liberal capitalism, and too anthropocentric—these critics argue—and it is only through bolder politics that authentic wilderness protection can be achieved. While it is not necessarily easy to distinguish among preservationism, biocentrism, ecocentrism, and deep ecology (and in actual ecological field practice these distinctions are often very blurry) it is possible to outline their central ideas and key theoretical points of disagreement. Let it be clear, however, that these movements are not distinct camps and that no orthodoxies exist within them. Their ideas exist along continua and it is very common for individual thinkers and practitioners to straddle across lines, borrow from different traditions, and disagree with one another internally. Our descriptions here are purely introductory tools which must be read as expedient simplifications.

Preservationism is an old idea, dating back to the conservationist movement. Preservationists, however, later took their distance from the conservationists following a rift between Pinchot and Muir. Muir and the preservationists argued that nature is an ecosystem, in contrast to the conservationists' more atomistic view (Allin, 2008; Oelschlaeger, 1991). In practice, this ecological view (which is now dominant across the environmental sciences) argues that humans are related to the environment organically, as an internally related component, yet they retain the ability to impair specific environments' normal functioning. Preservationism calls for humans to understand their ecological function and to abandon moral and political values grounded in economic utility (Allin, 2008). But while preservationists are certainly more radical than conservationists in their attitudes toward environmental protection, preservationists remain committed to a value-free science and to liberal democracy—thus drawing the criticism of ecocentrists, biocentrists, and the more radical deep ecologists.

Preservationism is also anthropocentric, according to biocentric and ecocentric critics, since preservationists believe that "human interests are the ultimate arbiters of value" (Oelschlaeger, 1991, p. 292). In contrast, biocentrism and ecocentrism more profoundly question the ideology of speciesism, or the idea that one particular species—such as humans—can remain at the top of a natural hierarchy (Berman & Lanza, 2010). Rather than human beings, life is what lies at the center of the system for biocentrists—indeed life *is* the system (Bogle, 2014). In different but related terms ecocentrists treat natural systems as the dominant force in the universe. Though biocentrism and ecocentrism are not synonymous, we have limited space here to tease out their philosophical disagreements and therefore we find it more productive to present them, instead, as a united front against resourcism and anthropocentric preservationism. Wilderness—biocentrists and ecocentrists argue—must be preserved not for its value for humans, but for its value in itself: for its value to life, including human life. Wilderness for them transcends any instrumental or narrowly calculated value. Because humans are not at the center of the universe their actions ought to be driven by the realization that they live in a natural community of equals and that the life of every member of that community is intrinsically valuable.

One of the most articulate and thorough manifestations of ecocentric and biocentric philosophy is deep ecology (see Devall & Sessions, 2001). Originally developed by Norwegian philosopher Arne Næss in 1973 (see Næss, Draengson, & Devall, 2009) but loosely inspired by the historical legacy of thinkers as diverse as Martin Heidegger, Baruch Spinoza, Aldo Leopold, and John Muir, and later developed by a host of social and natural scientists, deep ecology is one of the contemporary moment's most globally influential environmental philosophies. Deep ecologists share ecocentric and biocentric values in positing humans as equal members of a greater universal ecology, but go beyond most ecocentric and biocentric approaches by arguing for a radical reorganization of global society and a dismantling of modern industrial culture. In an early, seminal contribution Devall and Sessions (2001) outline the key tenets of deep ecology as follows:

1. The well-being and flourishing of human and non-human life on Earth have value in themselves (synonyms: intrinsic value, inherent value). These values are independent of the usefulness of the non-human world for human purposes.
2. Richness and diversity of life forms contribute to the realization of these values and are also values in themselves.
3. Humans have no right to reduce this richness and diversity except to satisfy vital human needs.
4. The flourishing of human life and cultures is compatible with a substantial decrease of the human population. The flourishing of non-human life requires such a decrease.
5. Present human interference with the non-human world is excessive, and the situation is rapidly worsening.
6. Policies must therefore be changed. These policies affect basic economic, technological, and ideological structures. The resulting state of affairs will be deeply different from the present.
7. The ideological change is mainly that of appreciating life quality (dwelling in situations of inherent value) rather than adhering to an increasingly higher standard of living. There will be a profound awareness of the difference between big and great.
8. Those who subscribe to the foregoing points have an obligation directly or indirectly to try to implement the necessary changes.

Deep ecologists' focus on human population control, simple living, recognition of cultural diversity and aboriginal ways of life, and wilderness preservation are undoubtedly radical, but also most certainly coherent. As Johns (1998) argues, our planet can only sustain a limited amount of life, and it is just common sense to arrive at the conclusion that the more biomass is composed of humans, or used by them, the less will be available for other species. Wilderness preservation, therefore, is an absolute necessity and no conservationist compromise will simply suffice. This radical proposition, as we will see shortly, has been subject of much controversy.

Now, to a great degree the resourcist bias of much conservationist wilderness protection is a consequence of the dynamics of legislation and bureaucratization. However, one particularly nefarious outcome of the historical evolution of wilderness protection—ethnocentric bias and injustice—would have been much more avoidable, critics argue (e.g. see Neumann, 2004; Rothenberg & Ulvaeus, 2001). What has come to be known as the racialized response to, or the "Third World" or even "Fourth World" critique of, wilderness conservation revolves around the critical race idea that the dominant concept of wilderness carries a pernicious ideological slant arising as a result of its Western origin. Callicott (2008, p. 357) explains:

> The English colonists called the new lands of North America and Australia "wilderness," an idea originally taken from the English translation of the Bible . . . . This designation enabled them to see the American and Australian continents as essentially empty of human beings, and thus available for immediate occupancy. The Australian bureaucratic term for wilderness, *terra nullius*, a Latin phrase meaning "empty land," says it all quite explicitly.

There was nothing empty about Australia, New Zealand, America, or Canada, however, and the "erasure" (Plumwood, 1998) of indigenous populations from their native land did not stop at first contact, but continued with the onset of American- and British-led efforts to protect wilderness worldwide (see Sandlos, 2008; Spence, 1999).

Box 2.2 TERRA NULLIUS

In Latin, the expression "terra nullius" refers literally to "nobody's land." Originating in Roman law the terra nullius clause has historically been used to justify the occupation of territories that were not under the sovereign control of any state. Nevertheless, the terra nullius clause was also utilized by British settlers in order to legally justify their occupation of Australia, in spite of their recognition of the presence of indigenous inhabitants. In doing so the colonizers essentially treated Australia as a wilderness: a vast space where only nature, not civilized humans, had shaped the land over the course of time.

European settlement of Australia began in 1788. At the time of the Europeans' arrival different groups of indigenous Australians throughout the island had different types of social organization and norms. Yet, because of their marked difference from the British system, settlers could not understand them and recognize them as formal institutions. British settlers were also unable to find leaders with the authority to sign treaties. For all of these reasons Australia—as opposed to other British colonies—was occupied without the signing of treaties and it remains treaty-free to this day.

In 1982 Australia witnessed an earth-shattering change in its legal system when Eddie Mabo and four other Torres Strait islanders began to file legal proceedings to gain the right to own their traditional land: Murray Island, or Mer. After ten years of legal battles the Queensland Supreme Court and Australia's High Court concluded that the claimants had indeed legally owned their land prior to the settlers' occupation of Australia. This ruling effectively ended the validity of the terra nullius clause.

But the end of a legal discourse is not synonymous with the disappearance of terra nullius-like claims in other social spheres. The idea that a land belongs to no one, and that it is therefore a more or less pristine wilderness where only the workings of "nature" can be observed, is still at the basis of many attempts worldwide to set aside land for conservation purposes. In Australia and elsewhere large conservation projects have been, and in some cases still are, gazetted without due authorization from or even consultation with their traditional owners. Often these cases involve marine conservation zones because European colonizers typically view sea and ocean spaces as "empty" of human presence and thus free for resource exploitation.

One of the earliest and most vocal "Third World" critics of the wilderness idea was Indian historian Ramachandra Guha, who in 1989 published a scathing sociological critique of deep ecology. While welcoming the transition from anthropocentrism to ecocentrism that deep ecology embraced, Guha (1998, p. 234) found deep ecologists' principle that "intervention in nature should be guided primarily by the need to preserve biotic integrity rather than by the needs of humans . . . unacceptable." Guha went on to provide evidence of what makes this principle politically dangerous, highlighting the case of Project Tiger—a network of Indian parks touted by the international environmentalist movement as a success in the efforts to save the tiger. While Project Tiger may have done much to protect endangered felines, Guha argued, it also resulted in the physical displacement of village residents trapped within newly formed park boundaries. And more broadly, Guha (1998, p. 235) argued, "because India is a long settled and densely populated country in which agrarian populations have a finely balanced relationship with nature"—unlike the US where the "preservationist/utilitarian division is seen as mirroring the conflict between 'people' and interests"—"the setting aside of wilderness has resulted in a direct transfer of resources from the poor to the rich."

The Indian or Australian cases (see Box 2.2) are only two amongst many. As we will see in Chapters Five and Six, wilderness protection and management are bound to create political waves of conflict. But in addition to colonial and postcolonial displacement, economic inequality, and social injustice the ethnocentric bias of wilderness can also be felt at a more ideological and discursive level. In Canada, recent studies, for example, have shown that "wilderness narratives g[i]ve primacy to whiteness as a defining trait of Canadian national identity" (Baldwin, 2010, p. 886) and that discourses associated with Canadian wilderness spaces have been found to marginalize the idea of cultural diversity (Sandilands, 1999). In the US cultural context, DeLuca and Demo (2008) have outlined the history of eurocentric racial bias and class bias present in the history of the American wilderness movement, while Smith (2008) and Johnson and Bowker (2008) have compensated for the white bias in wilderness historiography by tracing the neglected history of the idea of wilderness in, respectively, the literary and political writings of African Americans, and their collective memory.

DECONSTRUCTION AND RECONSTRUCTION

While it was not always easy to trace beginnings for the three periods we have described so far in this chapter—the birth of a historical era, after all, can hardly be a finite event—there should be no disagreement in pinpointing the onset of the wilderness deconstruction moment: 1996. That is the year of publication of *Uncommon Ground*, a volume edited by environmental historian William Cronon following a University of

California-sponsored interdisciplinary seminar on the theme of "Reinventing nature." Comprising 17 chapters written by a variety of geographers, historians, and cultural studies and environmental studies scholars, the book uniquely succeeded in stirring both public and academic debate, serving for some sympathetic readers as the definitive theoretical foundation in the constructivist study of nature, and standing—for some, less sympathetic, reviewers—as an all-out ontological declaration of war. The chief intent of the book, as already mentioned in Chapter One, was to show how wilderness and nature are products of human imagination and knowledge, and not natural realities existing outside of human influence.

Uncommon Ground was not particularly original in absolute theoretical terms. Social constructionism, or constructivism, had been an accepted theoretical tradition in the social and cultural sciences for at least two decades at the time. Constructionism was also very much aligned with the ideas of popular leaders of postmodernism and post-structuralism—theories that had been very influential for quite some time. Simply put, social constructionism is an ontology (a theory of reality) based on a relativist epistemology (or theory of knowing). Relativist epistemologies are based on the principle that what we know is based on *how* we come to generate, share, and interpret knowledge, and precisely how we do so depends on a large variety of social, political, economic, linguistic, geographic, and cultural factors. Knowledge, and therefore reality, are in sum not independent from, but profoundly dependent on, knowers and their ways of knowing. The corollary of all this for the contributors to *Uncommon Ground* was as clear as the morning sun: though there was a time that life existed on this planet before humans did, there is now no such thing as a nature that is clearly separate from culture. We, humans, know nature through the discursive tools we have created to understand it, and this could not be otherwise. The idea of a contemporary untouched and pristine nature, or wilderness, separate from human civilization, is therefore pure utopia. The trouble with the received idea of wilderness (not the places, the idea) is that it led us to naturalize its historical formation and also to forget that (1) wildness is everywhere around us and therefore that (2) environmentally responsible conduct should take place everywhere, not only in the most spectacular and sacred places we have come to define as wilderness—an argument that renowned environmental author Michael Pollan had already notably written about in 1991.

There was another problem with the idea of pristine wilderness, argued Plumwood (1998). Insofar as colonization was intended as a discovery of untouched land by male explorers, it was all too easy to conflate wilderness and nature with virginity and femininity (Merchant, 1996; Plumwood, 1998). And yet, as the virginity of land was broken through masculine conquest, nature ceased to be sacred. Moreover, allusions to broader gender dynamics—which Plumwood identified as necessary to deconstruct how we, as a Western society, had come to think of wilderness—did not end there. Wilderness was no place for a woman, argued Plumwood (1998; also see Schaffer, 1988), unless apparently escorted and kept safe by expert male guides. Wilderness was the testing ground for male virtue, for physical toughness, and brotherly camaraderie, and its domestication (and thus feminization) only spoiled it and rendered it weak and impure, dragging down with it supposedly natural male fortitude and spirit. In Plumwood's (1998, p. 661) words, "if feminine immanence is expressed in limitation to the domestic, masculine transcendence is expressed in escape to a superior realm of true spiritual or adventurous experience in a nature defined against an inferiorized, familiar sphere of dullness and dailiness."

It is easy to identify the most immediate casualties of this deconstructive rhetoric. From the male Romantic writers who sought salvation in wildness to the European colonizers who erased indigenous presence from the land in the name of civilization, and from the activists and political representatives who defended a natural reality by inventing it outright to the ecological managers and tourist operators who profited from conserving Eden, much like one would safeguard history by displaying exhibits in a museum, social constructionists and colleagues scorched the earth that wilderness defenders had struggled to keep safe for decades. Soon enough Cronon's (1996b) lead essay, "The trouble with wilderness," would become the subject of heated newspaper editorials, scathing magazine review essays, and controversy-filled journal issues. Cronon himself in the process became ridiculed by colleagues as a privileged armchair academic and a city slicker, and the target of enraged activists who accused him of failing to understand the most basic ecological principles.

But why had *Uncommon Ground* touched a raw nerve? According to Proctor (1998) the opposition to wilderness deconstruction was fueled by irreconcilable worldviews. "Postmodern deconstructionism," as the critique symbolized by *Uncommon Ground* derisively became to be known, was reputed by wilderness defenders to be as dangerous as chainsaws and tractors. Defenders of the traditional wilderness idea thought that the constructionist argument was ignorant of biological laws and ecological principles, full of misunderstandings of the wilderness movement, and guilty of neglect in foreseeing its political consequences. Even though Cronon published a handful of extremely coherent and insightful self-defense pieces such as a provocative piece on wilderness preservation as hybrid nature–culture heritage conservation (Cronon, 2008), the "deconstruction" moment was well under way, for better or for worse, and it was not until cool tempers prevailed after years of debate (see Callicott & Nelson, 1998; Nelson & Callicott, 2008) that a way forward presented itself.

The way forward, we believe, is for a "reconstruction" of the idea of wilderness based on more-than-human principles and concepts. We will expand on more-than-human geographies when we discuss more thoroughly the concept of wilderness as assemblage and meshwork in Chapter Seven, but for now let us situate such "reconstruction" in the historical moment we have been discussing. Right around the time that Cronon's volume was published, geographers, sociologists, anthropologists, and cultural studies scholars had begun winning key battles in the "war" against modernism. Nature could no longer be perceived as separate from culture, and concepts such as socionatures, technonatures, and naturecultures—which describe the inextricable relation between society, culture, nature, and technology—soon became widely accepted (see Castree & Nash, 2006; Swyngedouw, 1999; Whatmore, 2006). Whereas social constructionism still clung on to an idealist humanism, which for the most part relied on the anthropocentric idea that people "attach meaning" to their world as if it were a *tabula rasa*, posthumanism introduced an ecocentric ecology—or shall we say a "flat ontology"—in which humans no longer played the role of writers, directors, or protagonists.

Posthumanism, according to Castree and Nash (2006, p. 501), "names a contemporary context in which new scientific developments trouble the foundational figure of the human subject as distinct from other animal forms of life." However, while the advent of *new* technologies and scientific developments is often at the center of posthuman theorizing and research, the "newness of complex contemporary entanglements of people, animals, technologies, and things" (Castree & Nash, 2006, p. 501) should not distract us from the fact that the human subject was never sovereign over, or separate from, nature (Ingold, 2010). In the words of Bruno Latour (1993) we have *never* been modern (emphasis ours). The ontological separation of humans from nature, the disabling of nature's agency, and the relegation of the non-human world to a dependent variable of human activity—all quintessential components of the modernist worldview—were and have always been a "historical fallacy" (Whatmore, 1999, p. 10). A "post-natural" (see Anderson, 2009) conception of wilderness, therefore, should not be based on the wrongful premise that there could ever be or ever was a "natural" wilderness, or that now is the ideal time for a "post-natural" one because contemporary technologies and sociocultural realities are different from the past. Post-natural or post-human approaches to our subject matter, rather, simply ought to be based on theories and concepts which emphasize the inevitable relationality of all actors within a network, meshwork, or assemblage and which introduce "a shift from ontological stability to ontological instability" and therefore a worldview "where classification by discrete, fixed categories has given way to entities/processes" converging and diverging "in the constant remaking of the world" (Anderson, 2009, p. 121; also see Ingold, 2010).

An example of such an approach comes from the work of Whatmore and Thorne (1998). "Wilderness," for them, "is inextricably social" (1998, p. 437). Therefore "the futures of earth creatures (including humans)," Whatmore and Thorne (1998, p. 437) argue, "lie not in fortifying the utopian space/time of a pristine wilderness," but in paying attention to "the everyday worlds of people, plants and animals [that] are already in the process of being mixed up." Such, we agree, is a concept of the wild that treats wilderness not as a natural entity but instead as a "relational achievement spun between people and animals, plants and soils, documents and devices, in heterogeneous social networks [or better yet, we will argue, meshworks] that are performed in and through multiple places and fluid ecologies" (Whatmore & Thorne, 1998, p. 437). In the chapters to come we will refer to this relational achievement as assemblage and meshwork.

And in the meantime, outside of the rarefied halls of academia, new ideas have begun to reconcile the imperatives of preserving nature while simultaneously paying respect to cultural heritage. For example, Cowley and colleagues (Cowley *et al.*, 2011–2012, pp. 29–30) observe that:

> Cultural resources—archaeological sites, ethnographic resources, cultural landscapes, and historical structures and sites—are components

> of wilderness areas and may contribute positively to wilderness character. In addition to preserving ecosystems, wilderness helps us understand human use and value of the land over time. One of the fundamental purposes of cultural resources is to promote multiple views of history, and wilderness can also be valued from multiple viewpoints. For example, a wilderness trail may reflect centuries of use by hunters, traders, miners, settlers, and travelers; today this same trail is used by wilderness visitors and represents a merging of past and present. Ecologically, while past human presence may not be apparent on a landscape, "the legacies of historic land-use activities continue to influence the long-term composition, structure, and function of most ecosystems and landscapes for decades and centuries after the activity has ceased" (Wallington *et al.*, 2005; also see Foster *et al.*, 2003). All wilderness areas have a human history.

In the US, for example, this reconciliation is driven by agencies such as the National Park Service (NPS) National Wilderness Steering Committee (now the Wilderness Leadership Council), which aims for policies to "properly and accurately incorporate cultural resource stewardship requirements into the management standards for wilderness areas" (National Wilderness Steering Committee, 2002, p. 4). Similarly, at the world level, UNESCO's World Heritage Convention has introduced a third category of heritage, a "mixed" category of nature and culture, which intends to recognize and protect "cultural landscapes" (see Taylor & Lennon, 2011).

SUMMARY OF KEY POINTS

- Ideas of wilderness have changed across modern times in light of historical events and in relation to changing social, political, economic, and cultural processes. We have outlined four historical moments that have affected how different people worldwide understand and value wilderness. By surveying the thought and actions of globally renowned leaders, thinkers, and writers we have identified four key historical moments.
- The first moment we discussed was wilderness the sublime, a moment parallel with the evolution of the Romantic movement of the late

eighteenth and early nineteenth century. Wilderness the sublime is an overpowering, enlightening, stimulating, and inspiring experience capable of bringing us close to nature and God. Wilderness the sublime is the object of endless representations extolling its virtues, from poetry to landscape art, from philosophy to religious mysticism.

- The second moment pertains to the period between the end of the nineteenth century and 1964, when the US Wilderness Act became law. During this period leading to the formal definition of wilderness and its institutionalization into conservation laws around the world, many countries experienced a changing dynamic in the supply and demand of undeveloped land. Though undeveloped land was still abundant at the end of the nineteenth century, development pressures were starting to mount everywhere with the growth of the industrial economy, and as result of a new sensitivity toward nature new wilderness conservation philosophies became widely accepted and practiced.
- The third period is a period of contestation. Largely coinciding with the publication of Cronon's *Uncommon Ground* a growing number of intellectuals around the world began to question the idea of wilderness and its androcentric, ethnocentric, and anthropocentric ideologies. A heated debate ensued.
- The fourth moment is characterized by a re-appreciation of wilderness and nature from the standpoint of posthumanism and more-than-human geographies.

DISCUSSION QUESTIONS

1. Is wilderness the sublime still common in contemporary culture? In what ways?
2. Is the legacy of modernist thought still exercising influence on the way some individuals and groups think of wilderness?
3. Why has the popularity of Thoreau's writings increased over the years?
4. Are legal definitions of wilderness such as those of the US Wilderness Act and the IUCN beneficial for the welfare of wilderness?
5. What other factors besides those we listed in this chapter contributed to the legalization of wilderness worldwide?
6. In what ways are contemporary ideas of wilderness androcentric?

7. In what ways are contemporary ideas of wilderness ethnocentric?
8. In what ways are contemporary ideas of wilderness anthropocentric?
9. Do you believe that the negative reaction toward the ideas contained in *Uncommon Ground* was justified?

KEY READINGS

Callicott, J. B. & M. P. Nelson (eds) (1998). *The great new wilderness debate.* Athens, GA: University of Georgia Press.

Cronon, W. (ed.) (1996a). *Uncommon ground: Rethinking the human place in nature.* New York: W.W. Norton & Co.

Nelson, M. & J. B. Callicott (eds) (2008). *The wilderness debate rages on: Continuing the great new wilderness debate.* Athens, GA: University of Georgia Press.

Thoreau, H. D. (1962 [1854]). *Walden.* New York: Time Incorporated.

WEBSITES

The Thoreau Society: www.thoreausociety.org

The John Muir virtual exhibit of the Sierra Club: http://vault.sierraclub.org/john_muir_exhibit/

The Aldo Leopold Foundation: www.aldoleopold.org/home.shtml

The Wilderness Society: http://wilderness.org/

All of these websites were last accessed on December 2, 2015.

3

REPRESENTING WILDERNESS

Photographing turtles, Ecuador (Photo: April Vannini)

In May of 1992 a young man by the name of Christopher McCandless arrived at a remote area nearby Alaska's Denali National Park after a long and tumultuous soul-seeking journey from the American South. Upon arrival Chris—who liked to refer to himself as Alexander Supertramp—set up camp inside an old abandoned city bus. Keen on living off the land as a test of endurance and as a way of uncovering the meaning of life, over the coming months Chris/Alexander would struggle to survive in the Alaskan wilderness, eventually succumbing to death by poisoning-induced starvation. When found inside the bus by a hunting party in August of 1992 Chris's body weighed only 30 kilograms. His death would have seemed senseless, had it not been for the short but vivid notes he kept inside his diary and the photographs found undeveloped inside his camera. Little did Chris know during his final moments that Alexander Supertramp would eventually become a world-renowned icon and that the abandoned city bus he called home would turn into an unusual wilderness pilgrimage site for many of the countless people he had inspired.

Alexander Supertramp's notoriety began in January 1993 with the publication of Jon Krakauer's article on his death in the pages of *Outside* magazine. Then in 1996 Krakauer wrote a book on Alexander's life story, published under the title *Into the Wild*. The story's popularity exploded just over ten years later in 2007, when first a documentary (Ron Lamothe's *The Call of the Wild*) and then a popular major motion picture (Sean Penn's *Into the Wild*) were released. Chris's life continues to attract attention to this day. *Return to the Wild: The Chris McCandless Story* is a 2014 PBS documentary not only detailing his biography but also examining the reasons for his notoriety. As *Return to the Wild* argues, it is clear that far from being the misguided adventure of a romantic college graduate, Alexander Supertramp's journey into the wilderness has struck a deep chord with an entire generation.

Throughout this chapter we will reflect on popular representations of wilderness and the wild—depictions, such as *Into the Wild*, which at one point or another have become very popular and very meaningful. *Into the Wild* is not an odd case. Popular culture is full of rich wilderness narratives. From adventure documentaries chronicling the feats of modern-day explorers to fiction films centered on the horrors lurking inside scary dark forests, from glossy magazines promoting backcountry travel to advertisements for the latest Leave No Trace camping gear, and from exciting paperback travelogues to the classics and neo-classics of literature standing as the founding myths of a modern nation's collective identity, wilderness plays a central role in popular media imagery and narratives.

In fact, given the recent massive proliferation of wilderness-related stories, both fictional and non-fictional, it was quite difficult to determine which representations should deserve our attention in this chapter. We could have, for example, opted to discuss exclusively contemporary products: the very latest and the most popular images and narratives today. But such a strategy might have proven to have an insidiously short-lived shelf life. Alternatively we could have focused on more "classical" depictions: popular and "high" culture that has stood the test of time. Yet in doing so we would have inevitably run the risk of covering representations of wilderness that are already too well known by some and completely unknown by others. In order to avoid both problems we chose to focus our chapter less on unique texts and more on generic characteristics that are shared by popular representations across media and across genres. The chapter is therefore organized along the lines of common themes: eight central and easily recognizable "frames" (more on "frames" shortly) that give wilderness representations their most common meanings and that provide their consumers with quickly recognizable communicative resources. These are: wilderness as adventure playground, last frontier, sublime space, self-renewal ground, enchantment, fragile ecosystem, national symbol, and lost civilization. For each we provide select examples from the most contemporary popular culture as well as from older representations across media such as television, cinema, books, magazines, and advertising.

Because there most certainly is not as much academic literature available (especially within human geography) for the subject matter of this chapter as there is for all the other chapters of this book, we take the liberty of blending existing knowledge with our own original interpretive insights and arguments. What do we mean by "frames"? Simply, interpretive patterns (see Goffman, 1974; Lakoff, 2010). Goffman and Lakoff intend frames as cognitive maps: tools which people use to make sense of reality. In few and simple words a frame is something that helps us select key aspects of a particular reality by making that reality easy to recognize and understand. Different frames can coexist alongside one another, so that when a particular frame is utilized rather than a competing one, we immediately recognize it as such without losing sight of the fact that it is only one of several alternatives. Before we dive into our frames we want to make a quick note on why representations are important. Even though our approach to wilderness as assemblage can be said to belong to the school of thought known as non-representational theory, we do believe that discourses and representations play an important role in assemblage formation. Non-representational theories after all do not deny the impact of representations, but rather advocate that research should not simply stop at the study of representation alone.

WILDERNESS AS ADVENTURE PLAYGROUND

The first media frame we examine is that of wilderness as an adventure playground. As we will see in greater depth in Chapter Four, wilderness areas around the world are very often used as adventure settings by a variety of outdoor recreationists. Mountaineers, modern-day explorers, paddlers, backcountry skiers, and countless other (often "extreme") sportsmen and sportswomen regularly turn to the undeveloped character of wilderness and the challenges these environments provide as a testing ground for their ambitions, physical skills, and specialized knowledge. As a result of the popularity of these adventurous pursuits, year after year many television programs, movies, magazines, advertising campaigns, lines of consumer products (see Box 3.1), and books are dedicated to the exploits of wilderness adventurers.

Wilderness adventures make for ideal dramatic content because of their inherently suspenseful structure. A journey into the wilderness has a natural narrative beginning, middle, and end. Formats and content are often very similar across productions. During the beginning our "heroes" will prepare for their journey by gathering the necessary resources, acquiring the relevant information, planning their expedition, reviewing and defining their goals and motivations, and establishing relations with one another, emerging as leaders or followers. This process reveals individual characters and personal stakes—allowing an audience to come to terms with why adventure matters, for whom, and in what ways. As the protagonists' journey begins the first few challenges immediately arise. Readers and viewers are then drawn in by the open-ended nature of the quest: will the heroes be able to surmount their obstacles? Will their initial struggles force them to turn around and abandon their quest? Will they ever be the same as a result of their experiences? Typically, wilderness adventure stories—fictional or documentary—follow a process whereby the stakes are constantly raised. Challenges, in other words, become greater and greater, forcing the protagonists to come to terms with the many ensuing crises. This raises suspense and keeps readers and viewers glued to the page or the screen. Then at last an adventure comes to its natural conclusion. A peak is either climbed successfully or not, an expedition is either completed safely or not, and so on. The heroes rejoice (or plan to try again in the future) and the true meaning of their adventure becomes manifest.

Wilderness settings magnify the nature of a challenge because their remoteness and undeveloped character force adventurers to rely on almost nothing but themselves, far away from help, comforts, and conveniences. As well, a wilderness area's dramatically intense weather conditions may complicate a journey and add unforeseen challenges. Wildlife may also impede a quest: big predators may add an element of fear and menace and small animals (e.g. mosquitoes) may add an easily relatable element of discomfort. Wilderness settings do not just play the part of an antagonist, however. Beautiful landscapes may provide both protagonists and audiences with a source of sublime inspiration and may serve as reminders that despite the mightiest challenges "nature" still remains more powerful and more important than any personal ambition. The adventure playground frame makes wilderness not just an inert backdrop, therefore, but rather a fully developed actor of its own. Wilderness is also an actor whose integrity and safety are at times compromised. In the adventure documentary *180° South* (2010),[1] for example, the protagonists are keen not only on conquering a remote Patagonian peak but also on raising awareness about the potential threat of industrial development in the region. The combination of the adventure frame and the fragile ecosystem frame enacted by this film is indeed becoming more common across different media and nowadays adventurers increasingly come across as superhuman not only because of their skills and courage but also because of their environmental leadership and stewardship (of course wilderness adventurers are not always such good stewards; at times their exploits are indeed just exploitative).

An example of the wilderness as adventure playground frame comes from the 2014 short television series *Chasing Shackleton*.[2] The three-part documentary describes the unique re-enactment of one of the most amazing modern-day survival stories. In 1914 Sir Ernest Shackleton and his team sailed to Antarctica, attempting to be the first to reach the South Pole. But owing to a shipwreck Shackleton and crew were forced to endure a three-year survival odyssey which only ended after the legendary explorer managed to desperately sail a small dinghy across the Southern Ocean in search of rescue. In *Chasing Shackleton* celebrated explorer

Tim Jarvis and a small crew of some of the world's best-known modern adventurers retrace Shackleton's journey by re-enacting his expedition in period clothing and gear. The program follows their treacherous journey over both sea and mountains, as Jarvis and his team endure the same struggles experienced a century before by Shackleton and his men. Just as was the case in 1914, in 2014 the Antarctic wilderness mightily complicates every single aspect of the feat, as tall waves, strong winds, and glacial temperatures almost crush the resolve and the bodies of each member of the crew one by one. Yet in the end the will to survive triumphs for Jarvis and his team, as it did for Shackleton and his men.

Box 3.1 OUTDOOR RECREATION AND CONSUMERISM

As Buckley (2003) points out, there are minor conceptual distinctions between adventure tourism and adventure and outdoor recreation. Nonetheless, outdoor recreation, adventure recreation, and adventure tourism overlap significantly when it comes to the production, marketing, and selling of products such as gear and apparel. Gear and apparel may seem like mere luxury options, but in a consumer society it is difficult to think about participating in various recreational activities involving experiences in the wild without thinking about "appropriate" gear and apparel. As a result, recreational and leisure activities such as hiking, trekking, camping, climbing, ocean kayaking, white-water kayaking, windsurfing, surfing, snowboarding, skiing, caving, mountain biking, diving, and snorkeling have recently become a big industry for retailers of apparel and gear companies that specialize in marketing to the novice and weekend outdoor recreationists, as much as to experienced and professional adventurers.

Moreover, clothing and gear are often fashionable on their own, so it is not uncommon for many companies to market outdoor recreational fashion to all consumers, regardless of whether they even attempt to participate in any kind of outdoor activity. As Buckley (2003, p. 126) notes, "particular clothing companies use sponsored athletes and specialist lifestyle entertainment media to sell clothing and accessories at both a high volume and a high mark-up to nonsporting, but fashion-conscious urban consumers, and adventure tourism is one of the links in the marketing chain."

Consumerism is infused in almost all aspects of contemporary Western culture, outdoor recreation included. Consumerism is a characteristic of a culture that thrives on consuming not only material objects, but what we experience as well. Outdoor recreation companies commodify experience by selling products and marketing images and discourses that relate to an "authentic" outdoor recreation-centered lifestyle and by utilizing strategic communication that carefully constructs what counts as an "authentic" experience. These ideas of an "authentic" experience are routinely evoked by branding images of wilderness and the "great outdoors" through high-end fashion apparel and specialized gear. In this way merchandise companies and retailers produce a brand image that appeals to a specific consumer who participates in outdoor recreational activities and who displays a certain lifestyle and a particular taste associated with it.

Nowadays, it is not uncommon to find a large selection of wilderness-related equipment within stores dedicated to outdoor clothing and gear. Specific brands market to a growing population of wilderness enthusiasts, thus cultivating not only lifestyle identity but also a carefully constructed image of how one should engage in activities in the wild. Thus, along with the "right" wilderness high-tech safety gear, consumers are instructed to also create a desire for the "right" environmental fashion. In this way, retailers like Mountain Equipment Co-op and Patagonia in particular have developed into a progressively large and influential industry that promotes itself as doing the right thing. As Hepburn (2013) states, companies like Patagonia often brand their products as "ecofashion," appealing to a core customer who is:

> a very fit person who aspires to step outside mainstream society, and who engages in extreme sports through which they have transformative experiences in sublime wilderness

Continued

> landscapes. Patagonia's marketing strategies and business practices intentionally minimize environmental damage, promote sustainability, and encourage people to appreciate wilderness and what can be experienced in it.
>
> (p. 623)

Retailers sell not only merchandise, services, and gear, but also images and ideas of wilderness. Vander Kloet (2009), through a discourse analysis of Mountain Equipment Co-op (MEC) catalogues, argues that MEC not only sells outdoor consumer goods, but also constructs a definite image of wilderness and nation that creates a particular consumer subjectivity, that of the eco-consumer. Vander Kloet (2009) suggests:

> MEC grafts together a subjectivity which positions consumption as a satisfying means of political engagement. To some extent, the production of the MEC as a 'conscientious eco-consumer' pre-emptively squelches the possibility of further engagement with environmental concerns. This eco-consumer subject is enticing because of what it allows MEC members to erase from their collective conscience.
>
> (p. 232)

Vander Kloet (2009) further argues that retailers like MEC are devoted to constructing how consumers understand wilderness in order to create a desirable space for outdoor recreation and for the consumption of the products required for outdoor recreational activities. Along with a specific idea of wilderness, retailers and companies selling the outdoor recreational lifestyle also create an environmental aesthetic that is infused in the products that are sold.

Moreover, companies such as MEC and Patagonia strive to promote a pro-environmental consciousness. Thus, their products create the image not only of a consumer that participates in outdoor recreational activities, but also one that follows environmentally sustainable practices. In this way, Hepburn (2013) suggests, "sustainable" technology and ecological responsibility are closely associated. For example, Patagonia has led the way by changing perceptions of the sustainability of synthetic fibers, cultivating the image of environmentalism with high-tech performance gear. Therefore, environmental consciousness, wilderness, safety, and lifestyle become packaged together as a single, multi-faceted commodity.

WILDERNESS AS LAST FRONTIER

Hard done by the global economic downturn of the first decade of the new millennium, a team of unemployed and nearly desperate American men from Oregon set out for Alaska and the Yukon wilderness in search of gold. TV cameras follow their every step: from the initial struggles in setting up camp to the discovery of the first nuggets of gold, and from the explosion of interpersonal conflict as difficulties mount to the rekindling of their dreams as brand new mining machinery finally allows them to strike rich pay-off dirt. This is *Gold Rush*: a Discovery Channel reality TV series in its fifth season at the time of writing. *Gold Rush* is not the only one in its genre, to be sure. *Bering Sea Gold*, *Bering Sea Gold: Under the Ice*, *Ice Cold Gold*, and *Yukon Gold* exploit the same idea in the same style on different time slots with nearly identical results. And then there are the shows dedicated to brawny and ornery men dead set on harvesting expensive seafood: *Deadliest Catch*, *Lobster Wars*, *Wicked Tuna*. And the list could go on. The point? Whether a source of rich minerals or delectable foods, wild places nowadays rank amongst the most popular settings for "last frontier" TV programing. The bush and high seas are "the new Wild West."

Wilderness as last frontier is first and foremost a kind of denial: a denial of the tamed, structured, confining, predictable, domesticated, and overly controlling society and culture typical of the city and the lifestyle it allows. Wilderness as last frontier is posited as an alternative to this society: a nostalgic realm characterized by open spaces (geographical and symbolic), infinite possibilities, rich opportunities, and few or no limits other than those that are self-imposed. Wilderness as last frontier offers a sense of freedom, a nostalgic space and time, a longing for a way of life dedicated to the cultivation of time-tested skills and the enjoyment of simple and basic pleasures. The wilderness as last frontier frame does not necessarily entail a consumptive and exploitative relationship with wild places' resources, of course. In many popular culture representations the last frontier is a setting for the cultivation of sustainable relations with nature. Several novels, memoirs, and homesteading manuals offer inspiring examples of practices of voluntary simplicity, for instance: a lifestyle centered on the embrace of moderation, skill, thrift, the spirit of making do, local knowledge, self-reliance, ecological education, and the refusal of consumerism, superficiality, conceit, pretense, excess, and dependence on a system of distant providers.

Take, for example, *Becoming Wild*, a recent memoir by Nikki Van Schyndel (2014). *Becoming Wild* tells the story of a young urban woman leaving the city and setting out to live for a year in British Columbia's remote central coast. Armed with only the most basic tools, traditional backcountry skills, bush medicine knowledge, and lots of courage and determination Nikki and a friend gradually learn to harvest food, make shelter, weave baskets and clothes, and establish a new rapport with nature based on mutual discovery and respect. By suffering discomfort and deprivation (and at times even hunger and physical pain), but also by uncovering the beauty of simplicity, Nikki eventually opens her eyes to the superficiality of the modern condition and learns to fully trust nature and rely on herself and her instincts. The last frontier may seem utopian at first, but as Nikki's experiences show (and many other similar narratives also do) the "natural world" works as the best and only alternative to the degraded post-industrial and consumeristic modern condition. Wilderness as a last frontier thus stands as a romanticized myth and nostalgic evocation of halcyon days but also as a real lifestyle alternative to existential dissatisfaction with the present.

In the English-speaking Western world wilderness spaces such as the Canadian North, Alaska, and the Australian Outback are often chosen as last frontiers not only because of their relatively undeveloped character but also because of the vast openness of their landscapes, the presence of large predators, and the intensity and unpredictability of their climate. Frost (2010), for instance, writes that films set in the Australian Outback routinely attempt to reconcile the Outback's overt hostility with the attractiveness of its landscape and the alleged simplicity of the indigenous lifestyles it is home to. In many such films the Outback "is constructed as an unknown frontier . . . where there is scope for personal self-advancement, self-discovery, heroic achievement (of survival, exploration, and collection) and experience of supernatural forces" (Waitt, 1997, p. 50). In sum, the frame of wilderness as last frontier depends on representations that strike a balance between danger/risk and opportunity/freedom and on portraying nature as threatening but also benevolent. Wild nature is an inexhaustible source of challenges and adapting to it inevitably demands sacrifice, pain, and suffering. However, wild nature also regularly and abundantly provides, especially after a person has changed his or her ways, learned all the necessary skills, and accumulated enough traditional knowledge.

WILDERNESS AS SUBLIME SPACE

As already mentioned in Chapter Two, portrayals of wilderness as sublime places have been common throughout history. Contemporary popular media representations of wild places as sublime are no different from their earlier depictions in classical literature or painting (cf. Flad, 2009). The notion of the sublime as we recognize it today emerged during the late classical period but was later revitalized and expanded throughout the Romantic movement in European thought. Simply put, the sublime is a mysterious, elating experience arising from the awareness of beauty. Typically occurring in the presence of nature itself or before aesthetically pleasurable representations of natural scenes, the feeling of the sublime can comprise two different but interrelated feelings.

The first, which can be referred to as the Burkean sublime after the thought of Edmund Burke on the matter, is based on an aesthetic sensation that is terrifying in its sheer vastness and dark enormity (Gandy, 1996). The Burkean sublime, in other words, is awe-inspiring, unsettling, and destabilizing. The second kind of sublime is instead concerned with the limits of the human comprehension of the might of nature. This "Kantian" sublime—following Immanuel Kant's treatment of the subject matter—works to remind us that our imaginative power is finite, our judgment is limited, and our understanding of nature is impossibly restrained. The sublime therefore confronts us by provoking in us an experience of a power infinitely greater than ourselves, a power that gives us insight into the human mind and spirit, and therefore a power that alerts us to the existence of a transcendent order.

In contemporary popular culture, one of the most visible proponents of wilderness as sublime space is German film-maker Werner Herzog. According to Gandy (1996), wilderness landscapes are central throughout Herzog's filmography. "Herzog's presentation of individual struggle as part of an existential rebellion against nature itself," argues Gandy (1996, p. 3), extends his world "beyond a spiritualized romanticism to a more gloomy assessment of the human condition." Herzog's films are characterized by a focus on the physicality of nature and by humans' situatedness in it, and yet, simultaneously, their removal from it. Gandy (1996, p. 5) explains:

> Herzog explores the limits of representation and meaning through the use of the sublime to denote the threshold of human comprehension. This is conveyed cinematically by presenting the central characters as visionaries struggling against nature itself, often involved in hopeless quests and narratives with anticlimactic endings. In emphasizing both the incomprehensible and the frightening aspects of the sublime in combination with an emphasis on the individual aesthetic experience, Herzog's work embodies aspects of both the Kantian and Burkean romantic traditions. Yet his aesthetic vision reaches beyond romanticism and the striving for spiritual unity with nature to encompass existential readings of nature as something morally empty and indifferent towards human life. . . . The search for the wilderness of "first nature" is a recurring theme in Herzog's work, in his use of what he has termed "unembarrassed" landscapes to form the setting for his films: an unblemished natural backcloth against which his allegorical narratives can be played out. The idea of wilderness "as the repository of transcendent truth and ultimate reality" forms the metaphysical underpinning to the use of landscape in Herzog's work, often using the narrative of a journey to symbolize a descent into a primordial past. Herzog's depiction of nature can be characterized as premodern in the sense that nature is portrayed as an overwhelming and inimical force, in contrast with the twentieth-century concerns with the destruction of nature as a fragile biosphere.

Another very well-known visual artist who throughout his illustrious career has portrayed wilderness as sublime is American landscape photographer Ansel Adams. Adams's photographs of Western landscapes typically concentrate on towering mountains, deep valleys, and impregnable canyons, routinely portraying wilderness as spectacular but also awe-inspiring. As Giblett (2009) has argued, Adams is a modern-day John Muir: a monumental artistic icon as much as a key leader in the popularization of the environmental conservation agenda. Adams "did his best to translate [Muir's] reverence for nature into spectacular nature icons" finds Schama (1995, p. 9), presenting us with images of formidable peaks and sweeping vistas that beg to be associated with the otherworldly. Known as the most outstanding photographer of the American park landscape (paralleling in this sense documentary film-maker Ken Burns, whom we will examine later), Adams's sublime pictorial representations of national parks promote not only the value of magnificent scenery, but also the power of nation and its ideas of wealth. As Giblett (2009, p. 44) argues: "Adams' mountain photographs are also sublime in the sense that they sublimate the land into landscape, into aesthetic object, the depths of land and space into pictorial surface, and also into a commodity, as he was a professional photographer." Beyond Herzog and Adams, images of wilderness as sublime space abound in many movies and photography, and are especially popular in advertising as they are routinely used to aggrandize and fetishize commodities, tourist destinations, and services.

FINDING SELF-RENEWAL IN WILDERNESS

In the domain of popular culture wilderness is not unlike an island: a place seemingly disconnected from the rest of the world where one can travel to break ties from the past and reinvent oneself from scratch. Numerous films and books capitalize on this image of wilderness as a ground of self-renewal and self-reinvention. The most recent to capture showbiz headlines around the world is *Wild*. *Wild* (full title: *Wild: From Lost to Found on the Pacific Crest Trail*) first saw the light of day as a book. The book is the memoir of the then 26-year-old American Cheryl Strayed. After the death of her mother, the end of her marriage, and a drug-filled personal crisis Strayed decided to hike alone the 1,100-mile-long Pacific Crest Trail from California's Mojave Desert on to the Pacific Northwest as a way to refocus her life. Far from an expert trekker, Strayed tells the story of her journey as an experience filled with fears, anxieties, and difficulties as much as with warmth, humor, delight, and hope, and as she continues to walk in spite of the odds we begin to see how the wilderness heals her. Following the bestselling success of the book, *Wild* (2014) became a sensation at the box office as well. Starring Reese Witherspoon as Cheryl Strayed the film echoes the book in every respect, also paying homage to another famous story of wilderness as self-renewal: *Into the Wild*.

As was the case with Chris McCandless's own quest for personal salvation Cheryl Strayed's journey constructs the wilderness as a place where physical fatigue, discomfort, and the lack of modern conveniences can allow one to focus on the truly important meanings of life. Wilderness indeed seems to be able to regenerate the mind, heart, and body because it breaks people down and forces them to adapt and pick up the pieces. It also allows for the kind of deeper reflexivity—facilitated by slow-passing time and physical distance from one's world—that gives someone clarity of mind and perspective. As you may promptly recognize, this is far from a novel idea; Henry David Thoreau himself had piloted the formula in *Walden* and only a few modifications have been applied since then. Indeed since the days of Thoreau's experiment the self-renewal frame has been so powerful because it strikes at a few core dimensions of Western selfhood. Let us reflect on three of these.

First, wilderness offers the opportunity for self-renewal by confronting someone with the possibility of death. Death is always looming in the background within this discursive frame, whether by starvation, injury, accident (e.g. fall, drowning, etc.), or by encounter with large predators. But rather than killing a protagonist—as may very well be the case in real life—these popular representations often depict death as a waking call. By coming into close contact with the possibility of death the protagonist suddenly realizes that s/he is mortal and more importantly that s/he cares about life. The potential of death by wilderness, therefore, functions as a way of confronting a "natural" way of living and dying. Potential death by wilderness, in other words, functions as a catalyst for the protagonist to regain control of him-/herself and his/her life.

Second, wilderness instills confidence. Neither Chris McCandless nor Cheryl Strayed was a trained backcountry survival expert. The stories of both show plenty of mistakes and poor choices—some of them so maddening that outdoor enthusiasts might find it difficult to enjoy their stories—but also show their willingness to learn, as well as their enthusiasm, determination, and pride in learning. As a matter of fact it is precisely through their learning-by-doing, through their trials by error, and through their discoveries that the protagonists of these representations gain a new sense of personal confidence. Now, to be sure, their wilderness skills do matter in and of themselves, but what matters even more within this frame is the protagonists' own sense of growth and self-trust. It is by gaining a new respect for themselves as whole individuals that our heroes and heroines learn not only to adapt in the unfamiliar environs of the wilderness, but even to thrive and to enjoy their rapport with nature more authentically.

Finally, wilderness yields "spiritual" salvation and here in a way the present frame overlaps with two previously discussed frames: wilderness the sublime and wilderness as last frontier. As Gould (2005) writes with reference to homesteaders seeking a better way of life in self-sufficient homes and farms, a greater harmony with nature means a better rapport with God. Popular culture representations of wilderness as a self-renewal ground do not miss out on this. Whether God is intended in the conventional religious sense, or in a more pantheistic way, a foray into wilderness forces one to confront nature's daily and seasonal rhythms and the sanctity of all forms of life. Chris McCandless, for instance, finds it to be one of the greatest tragedies of his life that he is unable to quickly preserve the meat of a moose he killed and that much of that food (and therefore that very life) goes to waste. It is through encounters like these that the protagonists of popular stories informed by the self-renewal frame come to terms with the transcendence of life and thus they eventually manage to obtain personal regeneration and spiritual salvation.

ENCHANTING WILDERNESS

Popular culture regularly frames wilderness as a fascinating and delightful realm where amazing things constantly happen. This frame, which we believe *enchants* wilderness, constructs wilderness and wildlife as a captivating domain where the wandering human imagination is routinely captivated by things that seem marvelous and otherworldly and yet so real and so "natural." However, as critical textual studies reveal, the reality status of many of these enchanting media phenomena is quite dubious. Take wildlife documentaries, for example. Though wildlife documentaries are commonly believed to be faithful to reality and not fictional—after all, that is the unique promise of the documentary genre—their narrative characteristics are obviously very shaped by fictional conventions. Bagust (2008, p. 219) lists the well-known qualities of these so-called "blue chip" wildlife documentaries:

1. depiction of mega-fauna, especially large predators;
2. visual splendour and spectacular scenery;

3. dramatic narrative (the animals are often anthropomorphized);
4. absence of history and politics;
5. absence of people (except occasionally tribal, pre-industrial or "natural" people, including park rangers);
6. absence of explicit references to the scientific method and the process of filming.

Scott (2003, p. 31) argues that there are four additional characteristics typical of wildlife documentaries. First, these documentaries are narrated as factual and are closed to viewer interpretations—that is to say, they portray documented and incontrovertible empirical evidence, not drama and narrative. Second, and to reinforce the first point, they are narrated by an omniscient male voice who speaks in an authoritative, paternalistic, and didactic voice, as in the "voice of God." Third, the events portrayed in these films are presented as unadulterated and natural facts that are neither scripted nor manufactured through editing. And fourth, in spite of the previous point, a classical and epic-sounding musical score is used to underline the dramatic and enchanting nature of the images shown. All of these characteristics lend wildlife documentaries their appearance of timeless actuality.

An important semiotic study conducted by Bousé (2003) allows us to further dissect the qualities of wildlife documentaries in detail. While in actuality it is extremely rare (and also rather unsafe in most cases) to approach wild animals to see them up close and interacting with one another in meaningful ways, Bousé (2003) observes that wildlife documentaries commonly allow us to stare right into animals' eyes and observe them as they act toward one another in a highly cinematic way (e.g. see Video 1). This is achieved through painstaking telephoto lens shooting that is done with a clear intent, Bousé argues (2003; also see Bagust, 2008; Pierson, 2005), to manufacture a narrative and generate drama. As film and TV viewers we are accustomed to narrative and dramatic content—we expect no less of wildlife programing and are immediately turned off when we do not get what we expect. Yet animal life in actuality is not so cinematic and therefore its portrayal needs to actively frame it as such during both production and post-production.

Beside the use of telephoto lenses and time compression, one of the key tools to achieve the intended dramatic effect is the use of continuity editing. Continuity editing consists of juxtaposing images that have the potential to be related to one another. By the mere act of logical juxtaposition viewers infer a relation between different images and interpret them as necessarily tied together. Images may have been shot at completely different times and locations, yet when we see on screen something like a startled-looking gazelle suddenly standing on its feet and staring right of screen, followed by a shot of a lion ready to pounce left of screen, we viewers impute a relation between the two random shots, and suddenly continuity editing manages to convince us that the lion is in the mood for a snack and that the gazelle is ready to flee for its life. Just like that, drama is made. These pseudo-relationships, misconceptions, and false expectations are regularly created by continuity editing and by the frame imposed by wildlife documentaries, argues Bousé (2003). Even soundscapes are at times subject to hyper-real representation. For example, red-tailed hawk calls are known to have been used in films in lieu of authentic eagle calls because they sound more majestic and powerful than the eagles' high-pitched hiccupped whistling.

Close-ups and editing techniques borrowed from fictional film-making also achieve other results, Bousé (2003) explains. They establish the character of an animal, for example, even though individualizing and establishing an animal's identity may have been achieved by utilizing images of different but indistinguishable animals. Cinematic conventions also assign first names ("Bambi," "Dumbo," etc.) and establish the point of view of an animal and thus create a perspective that viewers can identify with. As well, cinematic techniques allow for the attributions of human feelings and emotions to animals, and these create a sense of intimacy and a feeling of identification among the viewers. Identification with animals is an especially powerful phenomenon. By relating to an animal not only can we express empathy toward it (which is not a bad thing, of course), but we also tend to anthropomorphize it, and interestingly anthropomorphism can carry with it powerful ideologies. For instance we may attribute characteristics of our society and culture to animal groups, characteristics that legitimate various human imbalances and structural inequalities by finding their correspondents in the animal world (Giroux, 1999). The editing of Disney's *March of the Penguins* (2005), for example, has been found to naturalize gender, sexual, and economic ideologies by projecting them on to the penguin world (Stephen, 2010). Similar arguments have been extended to much of the wildlife programing of the Discovery Channel (Pierson, 2005).

Video 1 FRAMING THE GALAPAGOS

Watch the video at: https://vimeo.com/115005267*

This short video, filmed during our trip to the Galapagos Islands in 2014, tells two different stories from the Galapagos archipelago. The first part shows us a familiar world: amazing images of wildlife and breath-taking landscapes which we have come to expect from the islands. Presented as the gaze of a lone traveler, and accompanied by soothing music and only minimal ambient soundscapes, this first frame follows the typical style of nature documentary representation. But everything changes three minutes and 30 seconds into the video, when the second frame makes its "ugly" appearance. Startled by the arrival of other travelers and a cruise ship this frame reveals another aspect of the Galapagos, a very unfamiliar one. As nerve-jarring industrial music, wind noise, and human voices disturb the ear, our eyes turn to the Galapagos that wildlife documentaries never show. Images of development, consumerism, and tourism are now revealed as the camera lens is pulled back and reframed. We can now understand how wilderness can be easily generated through a selective gaze.

* Last accessed December 2, 2015.

WILDERNESS AS FRAGILE ECOSYSTEM

Few objects of public discourse are as characterized by contradictions as wilderness and nature. After all, the very fact that we could list as many as eight competing and often mutually contradictory frames (and more could have been added with additional space) is evidence of the fact that wilderness and the broader subject of wild nature are heavily contested and subject to dramatic historical shifts in public perception (see Macnaghten & Urry, 1998). Raymond Williams (1983, p. 222) was clearly right when in his famous compendium of culture keywords he wrote that nature is "at once innocent, unprovided, sure, unsure, fruitful, destructive, a pure force, and tainted and cursed." Expanding on

Williams's argument, British environmental philosopher Kate Soper later explained:

> Nature is . . . represented as both savage and noble, polluted and wholesome, lewd and innocent, carnal and pure, chaotic and ordered. Conceived as a feminine principle, nature is equally lover, mother and virago: a source of sensual delight, a nurturing bosom, a site of treacherous and vindictive forces bent on retribution for her human violation. Sublime and pastoral, indifferent to human purposes and willing servant of them, nature awes as she consoles, strikes terror as she pacifies, presents herself as both the best of friends and the worst of foes.
>
> (1995, p. 71)

The frame that most clearly embodies many of these contradictions is that of wilderness as fragile ecosystem. As opposed to other frames which have been part and parcel of discourses about wilderness for centuries (such as notions of the sublime), the fragile ecosystem notion is relatively recent as it can be said to have been born together with the strong growth in public concern about environmental problems typical of the last decades of the twentieth century. The fragile ecosystem discourse frames wilderness and nature as "endangered" and subject to constant human abuse. Acid rain, pollution, the thinning of the ozone layer, deforestation, development, and global climate change have all surfaced at one point or another within public discourse as the chief sources of concern. But regardless of the actual problem of the day, popular culture representations have invariably portrayed wilderness and nature as the weak and defenseless victim of vicious attacks by an overpowering and insensitive perpetrator: humans. The notion of the fragile ecosystem thus constructs wilderness as an organism whose delicate balance is altered by the irresponsible actions of an outside agent who does not belong in nature and who should either keep away from its timeless rhythms and workings or dramatically amend its way of life in search of vaguely defined "sustainable" practices.

Popular representations of wilderness as fragile ecosystem are quite formulaic. Despite amazing imagery and powerful environmental messages their narrative formats are remarkably repetitive. Let us take for example the IMAX 3D film *To the Arctic* (2012). Directed by Greg MacGillivray, produced by IMAX and Warner Brothers, and narrated by Meryl

Streep, like many other wildlife films *To the Arctic* turns a group of animals (polar bears in this case) into the protagonists of a carefully edited environmental narrative. As Meryl Streep says: "Of all the truly wild places left on earth, none are as majestic as the Arctic." But the Arctic, and the polar bears that live there, are as fragile as they are majestic. Global climate change is making the Arctic a warmer place, and polar bears' ways of life are changing. It is up to us to do something to save the cute cubs and their resilient mother, we are told. And therein lies the format: first, majestic places and adorable animals make us fall in love with the wild. Then the threat of destruction of our beloved wilderness is foreshadowed. Crisis ensues. Finally, the resolution is presented, and the future is right in our hands. We humans are the culprits and the saviors.

Images of melting glaciers and warming polar ice worlds are popular ways to convey the notion that nature is as wild as it is fragile (Braasch, 2007). "Public images" like these, write Olson, Finnegan, and Hope (2008, p. 1), "work in ways that are rhetorical; that is, they function to persuade." Environmental visual rhetoric consists of producing certain images (instead of others), framing angles of sight and points of view, as well as post-producing and selecting images in order to suggest a particular ecological orientation to the world. Polar bear images, ever so abundant in popular culture, are used as the rhetorical symbol par excellence of a wilderness and wildlife struggling for survival in the battle against climate change (Cox, 2013; Flannery, 2005). Polar bears—despite their ferocious might—are thus portrayed as striving to cope with a changing environment, having to swim farther distances, starving, and even drowning far away from the melting icy shores. As Cox (2013, p. 71) explains, "within the context of images of melting ice and news of climate change, images of polar bears function as a visual condensation symbol": a vivid icon that strikes at an audience's most basic values and stirs their emotions, condensing on to itself a nexus of profound concerns, fears, and anxieties about a broader and more complex issue. The polar bears' weakness (in spite of their apparent might) therefore captures powerfully the broader frame of wilderness and nature as fragile, threatened, and invariably awaiting our rescue.[3]

WILDERNESS AND NATIONAL IDENTITY

Nationhood and wilderness have nearly always been in close rapport. A country's natural landscapes—and in particular its rural and remote landscapes—are believed to transcend history and politics, and therefore symbolize a timeless national space which embodies core national virtues and the essential qualities of the people who call the place home (see Kaufman, 1998; Zimmer, 1998). Wilderness landscapes are an especially powerful national symbol. Wilderness, thought to pre-date social organization, is routinely ascribed qualities that reveal a nation's foundational "nature" and epitomize its social and cultural norms, values, and ways of life. In this sense wilderness endows nationhood with a sense of immutability, unity, and continuity (Jazeel, 2005). This frame therefore constructs wilderness as a symbolic space that is the origin of national hegemonic ideologies and discourses which essentialize a collective being with a fixed identity: the nation.

No wilderness areas are invested with deeper nation-building symbolism than a country's national parks (see Frost & Hall, 2009). National parks combine together widely shared virtues and values which embody national identity not only in the mind of a country's citizens, but also throughout the rest of the world. National parks assemble notions of a common space, a collective history and memory, relations with nature and God, visions of the future, prideful national politics, and a collective self-image. For a powerful and well-recognized example of how the wild nature of national parks is constructed as a symbol of national identity we may look at Ken Burns's 2009 PBS documentary series *The National Parks: America's Best Idea*. The *National Parks* series consisted of six episodes for a total running time of 12 hours. A few clips are available online at the PBS website and the entire series is available on Netflix. For those without the access, *The National Parks* consists of a multitude of attractive landscape shots and found historical footage together with narrated commentary by the director, excerpts from classical American literature on nature, wilderness, freedom, and democracy, and interviews with various experts and authors.

Ken Burns's representation of the parks does not limit itself to exalting the sublime qualities of beautiful places; it explicitly connects them to nationhood and American identity. It is not just nature, "it is American nature," Bryant (2010, p. 476) argues in a critical reading of the series. Ken Burns's stated goal for the documentary leaves no room for misinterpretation: "to get to the heart of a deceptively simple question: Who are we? That is to say, who are those strange and complicated people who like to call themselves Americans?" His film, he says, intends to look at "the way in which the sheer physicality of this great continent has molded us as a people" (Bryant, 2010, p. 476). For Ken Burns, national parks therefore assume a mythical and foundational status: a spiritual glue that binds American people together. US national parks embody individuality and freedom as well as oneness with nature and a collectively owned, common land that stands for the American model of democracy (Bryant, 2010). This is quite a nationalist, patriotic overtone. Even the otherwise cool-headed environmental historian William Cronon seems unable to help himself in his interview aired throughout several episodes of the series: "This wild land is essential to who we are as a people," he says, and "the land embodies the nation. It is the place where we come to celebrate what it is to be an American" (see Bryant, 2010, p. 478).

Obviously, American representations of wilderness are not the only ones to fulfill the goals of nationalism and collective image-building. In Iceland (though many other similar cases worldwide could be cited) wilderness is used not only to express an image of nationhood, but also to promote the country to tourists across the world (Lund, 2013; Olafsdottir, 2013a; Sæþórsdóttir *et al.*, 2011). Strategically promoted as "Europe's last wilderness," popular media and marketers represent Iceland by evoking the wild natures of its coastlines and Highlands, for example through such slogans as "Iceland naturally," "Nature the way nature made it," and "Pure, natural, unspoiled." The "nationalization" of nature occurring in contemporary Iceland, however, is far from a recent tourism promotion gimmick, as it parallels the recent rediscovery of earlier Romantic poetry, folklore, and classical writings about Iceland's history and culture (Sæþórsdóttir *et al.*, 2011).

The intermingling of classical representations of wilderness landscapes and national identity is also present in similar forms in Australia, where landscape photographers have contributed to the formation of Australian national identity and pride with their focus on images of the bush, the Outback, and the desert (Giblett, 2007). "Landscape photography (including environmental and wilderness photography)," observes Giblett (2007, p. 342), "plays an important role in how we see and live in relation to the land . . . [and] it is related to developing an environmentally sustainable Australia." Thus, just as it is the case in Iceland, such photographic images are used to promote Australia not only as a prime tourist destination, but also a uniquely environmentally responsible country that has come to peace with its nature.

WILDERNESS AS LOST CIVILIZATION

Our final frame is a scary one, and we've left it for last to conclude our chapter in a creepy and sinister way that might trouble your sleep tonight. Imagine this scenario. You and your friends decide to go spend the weekend at your uncle Bob's cabin in the woods. You plan on doing a little fishing, a little hiking, and a little partying. Nothing special: just a weekend away from the city, way back into the thick, deep, dark forest up the river and way off the highway. The weekend starts out great: a few beers, lots of laughter, and nothing or nobody around to disturb you or to be disturbed by you. Or so you think. As you and your friends are swimming and splashing in the lake by your cabin suddenly you realize that your friend Katie is nowhere to be found. You last saw her swimming in the lake. Her clothes and shoes are still by the dock, but she is nowhere in sight. You start shouting her name. You think it's a stupid joke. Then, you start looking for her. Frustrated, after a while, you reluctantly decide to phone search and rescue. But as you pick up the phone you realize that the line is dead. So you head into the woods to look for her. Soon it gets dark. And scary. And from here on we will let you complete the story with your own imagination, or more likely with the narrative bits and pieces that you recall from any of the last ten horror movies you have seen which begin just like this. You get the point of this frame: wilderness is a frightening environment where wayward souls, uncivilized peoples, and ferocious beasts and monsters prey upon unwitting visitors.

From *The Blair Witch Project* to *The Shining*, and from *The Cabin in the Woods* to *Lost*, classical folklore and contemporary popular culture teem with tales, movies, horror paperback books, and TV series taking place in remote, wild settings—from high mountain ranges to desert islands and forests—where evil awaits. Sæþórsdóttir and colleagues (Sæþórsdóttir *et al.*, 2011) remind us that wilderness has been the subject of fear and anxiety for quite a long time. In Iceland the Highlands have long been believed to be the land of supernatural beings, outlaws, trolls, ghosts, and witches. Until access to the Highlands improved in recent years few people dared to enter and this only contributed to the mythologizing of the place. Though many creatures were more mythical than real, and arguably more useful as great narrative characters than anything else, many Icelanders did have firm beliefs in the existence of outlaws who called the Highlands home. Outlaws were allegedly escaped convicts who survived by stealing sheep and terrorizing local farmers. Actual sightings of these fallen men, however, were far and few in between the numerous hints of their existence. Over the years several expeditions were launched by brave (and hesitant) men to find them, but no one came back with definitive answers, and a few did not come back at all—thus further fueling the myths. Sæþórsdóttir *et al.* (2011, p. 260) observe:

> Such landscapes of fear have been integral to the Western understanding of wilderness. The image of the Highlands as a place of outlaws and mystical beings resembles in many ways the image of the wild forests in Europe in the Middle Ages and the Old English notion of "wildeor" in the story of Beowulf. In Iceland the legends of outlaws took the place of the "wild beasts" in making the Highlands "wild" just as the presence of Native Americans and dangerous animals contributed to the idea of wilderness for settlers in North America.

Indeed the wilderness as lost civilization has been especially prominent in North American popular culture. Murphy (2013) argues that the sense of horror prevalent in all the various representations that have made use of this frame arises from the experience of losing oneself in the wild, and with that comes the experience of coming across those who have been banned by civilized society or who could or would never be part of it. But this is not a simple fear of the unknown. As Cronon (1983, p. 10) has suggested, "in the wilderness the boundaries between human and non-human, between natural and supernatural, have always seemed less certain than elsewhere." Uncertainty and unfamiliarity, in Murphy's (2013) interpretation, are mixed with a profound historical legacy marked by fear of the unknown and prejudice toward the inhabitants of wooded areas. The wilderness is self-willed, reflects Murphy, and with that comes an animist attitude toward the place; a sense that it has an uncontrolled will of its own. "This feeling that the wilderness has a profound, and not always positive effect, upon those who reside there," writes Murphy (2013, p. 16), "is one informed not only by long-standing perceptions of wilderness itself, but also of the individuals and communities that exist within it or just outside its boundaries." Any travel into the wilderness, therefore, strikes at one of the most fundamental sources of anxiety of humankind: the fear of the unknown, the fear of what is different, and seemingly beyond control and reason.

SUMMARY OF KEY POINTS

- Wilderness plays a significant role in popular media imagery and narratives.

- We have identified eight central and easily recognizable "frames" that give wilderness representations common meanings and that provide their consumers with quickly recognizable communicative resources.
- By "frames" we simply refer to interpretive patterns (see Goffman, 1974). Goffman intended frames as cognitive maps: tools which people use to make sense of reality. In few and simple words a frame is something that helps us select key aspects of a particular reality by making that reality easy to recognize and understand.
- The eight frames we identified are wilderness as: adventure playground, last frontier, sublime space, self-renewal ground, enchantment, fragile ecosystem, national symbol, and lost civilization.

DISCUSSION QUESTIONS

1. Can you provide additional examples for each of the eight frames we have identified?
2. What additional frames could be listed beside the eight we have identified?
3. Are certain media more popular than others when it comes to representations of wilderness? Why do you think that is the case?
4. What do you think is the true appeal of wildlife documentaries?
5. Would sober and realistic representations of wilderness be as popular as their more "romantic" counterparts? Why or why not?
6. Discuss the representation of nature in national parks advertising for your country. Is it a fair and balanced representation?

KEY READINGS

Cox, R. (2013). *Environmental communication and the public sphere*. Third edition. Thousand Oaks, CA: SAGE.

Macnaghten, P. & J. Urry (1998). *Contested natures*. London: SAGE.

Soper, K. (1995). *What is nature?* Oxford: Blackwell.

WEBSITES

Into the Wild (film): www.youtube.com/watch?v=2LAuzT_x8Ek

Return to the Wild: The Chris McCandless Story: www.youtube.com/watch?v=ceareV2sOx0

180° South: www.youtube.com/watch?v=cWBz_pxYC0A
Chasing Shackleton: www.pbs.org/program/chasing-shackleton
Gold Rush: www.discovery.com/tv-shows/gold-rush
Werner Herzog: www.wernerherzog.com
The Ansel Adams Gallery: http://www.anseladams.com/
Wild (film): www.youtube.com/watch?v=tn2-GSqPyl0
March of the Penguins: www.youtube.com/watch?v=L7tWNwhSocE
To the Arctic: http://tothearctic.imax.com/
Ken Burns's *The National Parks*: www.pbs.org/nationalparks/watch-video/#642
The Cabin in the Woods: www.youtube.com/watch?v=NsIilFNNmkY

All of these websites were last accessed on December 2, 2015.

NOTES

1 www.youtube.com/watch?v=cWBz_pxYC0A (last accessed December 2, 2015).
2 www.pbs.org/program/chasing-shackleton/ (last accessed December 2, 2015).
3 Of course we are not opposed to the cause of utilizing such documentary films to persuade viewers to take action against melting ice or dying polar bears. Much more modestly, and neutrally, we are identifying a common frame of portraying these problems, and indeed a type of environmental frame that is used to make conservation more recognizable and thus politically powerful (see Lakoff, 2010).

4

EXPERIENCING AND PRACTICING WILDERNESS

Camping, Canada (Photo: April Vannini)

Imagine being stuck in traffic—on a bus or in your vehicle—on your way back home from work or school. It is late winter and while days are still dark, you can already see the first signs of spring. Waiting for what feels like forever at a red light your mind escapes to the distant view of mountains. You dream of a weekend in nature, away from the noise and pollution of the city. You fantasize a fun-filled escape to the forest, to a placid lake, to the peace and quiet of the wilderness at the edges of the city. It is in this "romantic" way that wilderness becomes a liminal zone: a counterhegemonic (yet, an increasingly popular and often commodified) arena for the experience and practice of recreation, rejuvenation, and spiritual and psychological relief. Such is the focus of this chapter.

We cannot begin to gain a true understanding of the sustained appeal of wilderness as a place of respite from human civilization—perhaps even a site of resistance from its perversions—without an understanding of Romanticism. For most of the past three centuries Romanticism—intended as an ideological and cultural movement focused on the valorization of emotion, nature, originality, individuality, and passion—has fueled many misgivings about industrialization, urbanization, and the cult of reason. Romanticism in this sense can be understood as a rejection of the Enlightenment: a critical response to the oppressive conformism and the rationalism of the Age of Reason, a rationalism which somehow denied the more "basic truths" contained in raw nature. Romanticism has served as inspiration for both the artistic and the mundane appreciation and exploration of nature, and has provided our culture with the philosophical foundations for the exploration, contemplation, and conservation of nature. It is in the ideas generated by Romantic philosophers, writers, and artists that we can begin to trace the contemporary idealization of wilderness as a place for the cultivation of freedom, adventure, peace, relaxation, health, spiritual communion, leisure, and the sublime (Meyer-Arendt, 2004; Olafsdottir, 2013a). And it is in Romantic ideals (e.g. see Rousseau, 2004 [1782]; Thoreau, 1992 [1862]) that we can trace the contemporary view of wilderness as a prime cultural site for the rediscovery of what is truly important and meaningful in life. It is in Romanticism, in sum, that we can begin to acquire the foundations of much of the knowledge contained in this chapter.

Few Romantic thinkers would have made too much of a fuss about the distinction between the broad idea of "nature" and a strictly defined concept of "wilderness." Whether one seeks and finds wildness in canonically wild areas or in the stream running down the hill just outside of town, the Romanticism-inspired perception of wildness is variable and infinitely nuanced. Therefore in this chapter we make liberal use of empirical research and theory on nature-based experiences and practices across a wide variety of outdoor environments. Keen on understanding the embodied interaction between humans, animals, and the environment we turn to scholarship that has attempted to capture the ineffable nature of the affective experience of wildness, ranging from encountering (and killing) animals to camping under the night sky, from climbing mountains to canoeing on white waters, from collapsing due to pain and fatigue to rebuilding oneself along a trail. "Each experience" of the wild "has a story," observes Kahn (1999, p. xxiv): "It has meaning to us, because we're functioning and thriving in relation with wild nature rather than reifying it like a mausoleum." So it is stories, passed on through human-subject research, which we will catalog and organize in this chapter.

Though the subject matter of this chapter is vast and diverse, we will not stray away from our two central topics: practices and experiences. By *experience* we refer to felt perception, and more precisely to the ways in which wilderness environments are corporeally sensed, known, observed, and engaged with. Experiences are also a matter of moods, intentions, goals, and emotions (see Ingold, 2000). For the most part experiences of the wilderness, research tells us, are positive. Environmental psychologists explain that being in "natural" environments affects the nervous system positively by resulting in stress reduction and in restoration of attention. Natural environments are also experienced positively by most people because they typically lack negative sources of stress such as noise, pollution, and traffic—all urban phenomena which have been linked to sleep disturbances, increased stress levels, high blood pressure, and a variety of mental health disorders. Furthermore, natural environments such as wilderness spaces may facilitate and encourage psychologically and physically beneficial activities such as exercise, recreation, and interaction with significant others (see MacKerron & Mourato, 2013 for a review of all this). Nonetheless, in a few documented cases wilderness experiences may also be negative, especially when associated with fear, danger, lack of control, reduced comfort, or even terror (see Coble *et al.*, 2003; Koole & Van Den Berg, 2005). In the pages to come we will explore a wide variety of experiences of wilderness gleaned by different people in varied contexts.

By *practice* we refer to activities, actions, and undertakings of various kinds. Experiences and practices cannot be easily—or perhaps at all—separated. Hiking, for example, is an activity, but as a practice it is inseparable from the sensations of the feet touching the ground (as well as one's hiking boots, see Michael, 2000), the sounds of the forest, the sights of the landscape (Edensor, 2000; Ingold & Kurttila, 2000; Macnaghten & Urry, 2000;

Wylie, 2005), or the emotions it generates. A focus on practice, as much as on experience, is therefore a way of paying simultaneous attention to what people do and to how they experience a place. It is a way of understanding the formation of *lay geographies*: "everyday activities" through which an "individual works and reworks, figures and refigures an account of a place" (Crouch, 2000, p. 65). Our attention to practice is not unique, of course. Contemporary geographies as of late have witnessed a tremendous explosion in the collective interest toward people "doing" place. Such "performative" geographies shift our attention from talking about or imagining the wilderness to a concern with material and embodied specificities and their multiple relations with one another in actually lived wild places (Abram & Lien, 2011).

The study of wilderness experiences has had a long history, especially in relation to wilderness preservation. As Cole and Williams (2012) explain, wilderness spaces were distinctly different from other recreational spaces and thus it is important to study visitor experiences for the purpose of preservation:

> Part of the wilderness idea was to promote a new relationship between people and land, both in how wilderness lands were to be managed and in the experiences people might receive from wilderness visits. These experiences, the immediate thoughts, emotions and feelings associated with being in wilderness and the more enduring changes in attitudes, perceptions, and sense of self that arise from these encounters with wilderness, were considered likely to be unique and different from experiences in other recreational settings.
>
> (p. 3)

Cole and Williams (2012) provide a historical sketch and list various approaches which scholars have taken to studying wilderness experience. The authors outline three types: motivation-based, experience-based, and relationship-based. Finding inspiration in this approach, but also departing from it, we begin this chapter with an extended look at outdoor recreation and adventure education, reflecting in detail on the meanings of risk and skill. Subsequently we move on to wildlife encounters, both in the non-consumptive context of nature-based tourism and in the consumptive context of hunting and fishing in the wild. After that we turn to the third major focus of this chapter: the role played by wilderness in facilitating physical, spiritual, and mental wellness. Along the way we take some time out from the linearity of our survey to pay in-depth attention through three text boxes to the practice of camping (Box 4.1), to home making in the wilderness (Box 4.2) and to the gendering of trekking (Box 4.3).

Box 4.1 WILDERNESS CAMPING

Wilderness camping is a practice common to a wide variety of wilderness users. From professional mountain climbers to commercial gold seekers, and from scientific teams of researchers to vacationing families, wilderness camping appeals to (or conversely, may be a necessity for) many different individuals and groups with diverse goals and interests. Regardless of who does it and why, wilderness camping has a few key characteristics that distinguish it from the more "domesticated" version of camping practiced in RV/caravan parks and commercial or public campgrounds. Wilderness camping may be understood as a type of camping occurring in temporary shelters (e.g. tents) inaccessible by paved roads. But what precisely constitutes the practice of "camping"?

Leisure camps—as opposed to, for example, work camps, refugee camps, therapy camps, etc.—are liminoid spaces. They are interruptions, pauses, breaks: they stand in between the ordinary times we reserve for work and domestic life. Thus, camping "brackets" one's regular life; camping is an ephemeral pause in the rhythms of everyday life, a change of pace and place, a break from the structure of routine and habit, and an escape that allows us to experience time as duration and organic flow (Hailey, 2008). Yet, as fleeting as camping trips may be, camping is also an activity that may endure over time as groups of individuals and families may return to the same location year after year (Hailey, 2009; Kearns & Fagan, 2014).

Continued

Basic (as opposed to "luxury" campgrounds and camping "resorts") campsites that offer neither electricity nor running water or flush toilets demand that campers become responsible for making their own heat, light, and food, as well as for collecting water and for disposing of their own bodily waste. Wilderness camping raises the stakes even higher by creating the conditions whereby campers may need to bathe in cold creeks, be on the lookout for wild predators, procure firewood, and purify their drinking water. Wilderness camping—as opposed to the basic camping that may go on in a family-friendly national park campground—also requires walking (or snowshoeing, canoeing, kayaking, etc.) into a site, and therefore being solely responsible for physically carrying in and out all the necessary equipment.

Wilderness camping also requires that the utmost attention be paid to ensuring that camping sites show no damage as a result of the camping activity. The ethics of "leave no trace" or "no trace camping" demand that campers follow key principles to minimize signs of their presence. Principles include minimizing campfire impacts, camping on durable surfaces, leaving what they find, respecting wildlife, packing everything out which they brought in, planning carefully ahead, disposing of waste properly, and being considerate of others.

In North America camping is a "widely ramified institution linked to national character" (Brereton, 2010, p. 12; also see Kearns & Fagan, 2014, in regard to New Zealand). Almost all of us live in houses or flats over which we have limited or no control. Water flows from the tap, electricity works by the flick of a switch, food is available in the refrigerator (or at the grocery store and take-out restaurant), heat is centrally distributed, and waste is quickly flushed away. Camping unsettles this way of dwelling in the world. By forcing us to become involved in our own survival—even in the short amount of time for which it is practiced—camping demands we become entangled with place and weather. And so by taking charge of a small site and reinventing how we bring comfort to our shelter, we play and experiment with a different sense of daily order, temporarily reimagining life and our own identity (Kropp, 2009).

Camping is also a ritual with a recurring structure: "we leave home, we arrive at a site, we clear [or clean] an area, we make and then finally break camp before departing" (Hailey, 2008, p. 2). Camping is then a dual process of deterritorialization and reterritorialization (Hailey, 2008). Deterritorialization is a process of taking away control and order from a place or territory. This is generally followed by reterritorialization: a process whereby ties are reorganized, reassembled, and subject once again to new forms of control and order. Camping is therefore generative of space—camping *makes* place (Hailey, 2008).

WILDERNESS EXPERIENCES AND PRACTICES IN OUTDOOR RECREATION AND ADVENTURE EDUCATION

Outdoor recreation is an activity broadly encompassing all forms of leisure taking place outside of buildings and therefore inclusive of all kinds of interaction that people have in their "free time" with their natural environments. The study of outdoor recreation is a vast and dynamic enterprise that spans the disciplines of education, tourism studies, business, environmental studies, geography, and psychology and it is not our intent to summarize its key principles and theories here. What we are interested in—especially in this section and chapter—are personal *experiences* of the wilderness occurring during *practices* of outdoor recreation, what we might call experiences and practices of "wilderness recreation." Wilderness recreation, simply put, consists of "voluntary non-work activity that is organized for the attainment of personal and social benefit including restoration and social cohesion" (Kelly, 1996, p. 27) and that takes place in areas that participants consider to be wilderness or wilderness-like.

Wilderness recreation activities are simultaneously focused on providing participants with opportunities to have fun and to learn about the natural environment. According to Donaldson and Donaldson (1958, p. 17, original emphasis) "outdoor education is education in, *about*, and *for* the outdoors"—therefore it is no accident that most (if not all) of its prominent organizations make environmental education and conservation a key priority as well. Table 4.1 lists four of the best-known outdoor recreation and education organizations, along with some key characteristics

Table 4.1 Prominent outdoor recreation organizations worldwide

Name	Key founder	Since	Focus	Members
Sierra Club	John Muir	1892	Conservation	2,100,000
Scout Movement	Robert Baden-Powell	1907	Youth education	32,000,000
Outward Bound	Kurt Hahn	1941	Outdoor education	200,000
National Outdoor Leadership School	Paul Petzoldt	1965	Outdoor education	120,000

for each one. All of these organizations, and many more like them, aim to promote leisure *and* foster appreciation and understanding of basic biological and physical concepts and facts about the natural environment, the development of personal skills, and the pursuit of healthful and physical activity. The benefits of their activities for participants have been found to include self-realization, awareness and respect for the natural environment, effective communication and leadership, creativity and inspiration, rejuvenation, spiritual development, and greater abstract and bodily knowledge of the lifeworld (see Plummer, 2009).

Adventure

Wilderness recreation provides participants with a sense of adventure. The practice of adventure has been part and parcel of human civilization for millennia (one only need think of Homer's *Odyssey*, the story of a solo journey to the planet's most remote lands). Adventure's appeal for many people resides largely in the sense of delight experienced in exploring and confronting the unexpected, which is believed to be a survival mechanism (Csikszentmihalyi, 1990). Adventure provides its pursuers with elements of real or perceived risk, danger, difficulty, uncertainty, and the potential for loss of life and property (Berry & Hodgson, 2011) and therefore with qualities of an experience that is all but the antithesis of the modern life condition in its penchant for safety, control, planning, diffused responsibility, and access to information. Wilderness, in particular, makes for a great setting for adventure because of its undeveloped character and the opportunity to come to terms with basic survival skills and to encounter wildlife and, together with that, all sorts of peril of bodily harm.

Wilderness adventures in the context of recreation have been known to give participants the possibility to experience the intrinsically rewarding feeling of "flow": "the holistic sensation present when we act with total involvement" (Csikszentmihalyi, 1975, p. 58). Flow, Csikszentmihalyi continues,

> is a kind of feeling after which one nostalgically says: "that was fun," or "that was enjoyable." It is the state in which action follows upon action according to an internal logic which seems to need no conscious intervention on our part. We experience it as a unified flowing from one moment to the next in which we are in control of our actions, and in which there is little distinction between self and environment; between stimulus and response; or between past, present, and future.
>
> (p. 58)

The sense of flow inherent in wilderness adventure is characterized by the enjoyment of the patterns of the adventure activities themselves, the use of skills, the opportunity for developing friendships and a sense of companionship, the feeling of accomplishment, the pursuit of competition, the feeling of emotional release, and even the sense of social prestige and regard one acquires from completing an adventure (Csikszentmihalyi & Csikszentmihalyi, 1999, p. 157).

Recent research studies and theoretical models focused on the social psychology of adventure have indeed shown that there are many benefits associated with the sense of flow that wilderness settings and activities afford participants. Dustin, Bricker, and Schwab (2009) argue that outdoor recreation has significant health benefits for its practitioners such as increased physical fitness and reduced stress. In his extensive review Mullins (2014) cites benefits such as higher self-efficacy, self-actualization, and improved quality of life, skill development, self-expression, accomplishment, and more harmonious relations with the environment and co-participants, which accumulate through the natural tension inherent in adventure. In particular he argues that practices focused on encountering risk and uncertain outcomes:

> stem from the abilities of people to *negotiate* dynamic, supportive, and forceful environments. These abilities and environments are learned over time, along routes, and with others as participants

> within dynamic surroundings, be they urban, rural, or wild. These negotiations *are* skilled performances; they are uncertain, guided by particular values, and communicated using certain language; they employ equipment and occur in relation to specific environmental features.
>
> (Mullins, 2014, p. 133, original emphasis)

Confronting wilderness challenges and mastering related skills is an experience that is often transferred to other domains of life, thus facilitating multiple forms of personal empowerment (e.g. see Fischer & Attah, 2001; Kalisch *et al.*, 2011; Russell & Walsh, 2011), as we will see in greater depth in this chapter's sections on therapy and spirituality and self-discovery.

How outdoor recreationists manage risk in the wilderness has been and continues to be the subject of much research and debate. With regard to walking Solnit (2001), for example, has observed that challenge-driven competitive goals have been gradually replacing Romantic ideals about the appreciation of nature. Wilderness settings, therefore, are often treated as arenas for pushing the envelope of human capability (one has only to think of the constant search for oddball records on Mt. Everest) or for the execution of more mundane personal challenges—ranging from peak bagging in Great Britain (see Video 2) to running marathons in the desert. Dangerous pleasure adds value to life because its pursuit entails confronting the possibility of incurring death in a type of play that ends up transforming the meaning of life itself (e.g. see Abramson & Fletcher, 2007, in relation to rock climbing). Nonetheless, there is often very little chance in this sort of play. Far from being mindless adrenalin junkies involved in a Russian roulette, adventure seekers regularly collect useful information and follow best practices before undertaking an adventure, thus seeking to minimize uncertainty without completely extinguishing the potential for challenge, pain, fatigue, and danger (Abramson & Fletcher, 2007). Rather than embracing risk for the sake of risk, therefore, adventurers are mostly keen on confronting hazard as a test of willful action, knowledge, ability, preparation, self-confidence, and skill (Laviolette, 2007). On the other hand we must note that adventure pursuits have been at times criticized for prioritizing goal achievement over all other responsibilities, including acquiring knowledge of place (see Video 2; Wattchow & Brown, 2011).[1]

Box 4.2 LIVING IN THE WILDERNESS

Humans are ideally supposed to be visitors who do not remain in the wilderness—at least if we go by the US Wilderness Act. This notion has been subject to a great deal of criticism because it ignores the traditional practices and land claims of both indigenous populations and settlers who have made wilderness their place of dwelling and because it may result in evictions and displacements of wilderness residents following the legal designation of a tract of land as park or reserve. So can anyone, nowadays, up and move into the wilderness? How can one manage to live full-time in the wilderness? And is a wilderness area still wild after a human has chosen to remain there?

I (Phillip) spent two years between 2011 and 2013 traveling across all of Canada's provinces and territories to find people who live off-the-grid (i.e. in homes disconnected from electricity and natural gas networks) and document their way of life. Many of the off-gridders I found lived in exurban, rural, and peripheral areas. But in February of 2013 I met a couple who lived full-time far more remotely than anyone else and who indeed lived farther away from human civilization than anyone I had ever heard of. Ron and Johanna lived in Northern Saskatchewan, 100 km away from the nearest road of any kind, and 160 km away from the nearest town. They had lived there for about a decade.

Ron and Johanna's home could only be accessed by small planes equipped for landing on water or ice. The couple typically chartered that plane—a round trip would cost somewhere around $3,000—only twice a year to stock up on the few supplies they could not otherwise grow or build on their own. Their home was entirely self-sufficient for electricity and heat thanks to a hybrid system that made use of solar and wind (with a small back-up gasoline generator) to generate power, and locally harvested firewood for heating and cooking. Water was sourced from a well. Waste and sewage were sustainably disposed of locally. And while they bought a good deal of meat and other provisions during their

Continued

twice-yearly trips they managed to grow much of their calorie intake right in their garden and green house. Home-making and building skills allowed them to build, repair, and craft a good deal of the objects and technologies they needed on a daily basis. Despite all this, Ron and Johanna were no hermits: they kept connected to the rest of the world via a satellite link that permitted them to watch television, access the Internet, and make (emergency) phone calls.

What brought Ron and Johanna from their previous home in New England so far away from the rest of society? Simply put: a thirst for peace, quiet, and tranquillity; a sense of challenge and adventure; and the will to be as self-sufficient as possible. While their remoteness was unusually dramatic their values are common to many people who call "the bush" home across much of Canada, the US, and Australia. Getting away from it all is indeed an increasingly sought-after form of lifestyle migration, driven—unlike many migratory patterns—not by economic reasons but rather by the existential need to live life on one's own terms, in a more basic and simple (though obviously at times very complicated!) manner. And wilderness areas, of course, serve as prime testing grounds for such lifestyle experiments, allowing individuals to start fresh, to clean the slate—as it were—and reinvent modern living.

Did Ron and Johanna undo their wilderness in virtue of their own move? In other words, did they—in light of choosing to remain rather than simply visit and leave—turn their wilderness into a less pristine, less wild environment? One could very well argue that their cozy home—with its stocked pantry and warm rooms—wasn't wild. As a permanent structure meant to give its dwellers comfort and convenience their off-grid home wasn't much different from your home or ours. But just a few steps away from their wind turbine the bush was as wild as it would have been before, or without, their arrival. As Ron and I would walk the trail surrounding their land—as we did daily for the time I was their guest—we would quickly and punctually lose sight and sound of anything that wasn't wild. Meeting them taught me that choosing to remain in a wilderness area, and practicing a low-impact lifestyle, does not spoil the experience of the wild. But an important question remains: what would happen if more families moved to the same area? Would their arrival undo the wildness of the place? At what point does too much human residence spoil a sense of being in the wilderness? And how much human presence would be too much?

A good tool for understanding how adventure seekers manage risk and tackle challenges while traveling through wilderness environments is the concept of "wayfinding" (Ingold, 2000). Wayfinding is a type of improvised, learn-as-you-go, exploratory movement that "depends upon the attunement of the traveller's movements in response to the movements, in his or her surroundings, of other people, animals, the wind, celestial bodies, and so on" (Ingold, 2000, p. 242). Wayfinding is not just about finding your way, of course, but more broadly about drawing upon past experience and local knowledge to tackle challenges "on the fly." Mullins (2009) provides us with a good demonstration of this concept at work through his ethnographic study of a 100-day-long canoe expedition from the foothills of the Canadian Rockies onward to the Arctic Ocean, 2,680 km away. Throughout the journey he and the other paddlers learned, for instance, to understand how to synchronize their rhythms with the weather, the landscape, and each other, as well as how to feel the landscape as they went—like distinguishing from a distance "the sounds of a rapid from the confluence of a creek." Mullins observes:

> The river "speaks" in meaningful ways to the canoeist, who can then respond. Understanding and recognizing the affordances of canoe travel allowed the group to find narratives of past movement and glean meaning from the landscape. Like the paddlers we followed, we could determine safe places to drag our canoes, for example. Wayfinding helped us recognize our situation in the temporality of the landscape.
>
> (2009, p. 250)

The idea of wayfinding also prompts us to pay attention to the sensuous dimensions of the corporeal experience of wilderness adventures. Whereas earlier research on outdoor recreation and environmental perception had focused primarily on the visuality of landscape, current research has begun to unearth more complex sensory characteristics of wilderness practices and experiences. In her fieldwork among mountaineers in Scotland, Lund (2005), for example, reflects on the ways in which walking and climbing as kinetic and tactile practices directly contribute to the formation of landscape views for each individual climber. But a mountaineer getting to know the landscape, Lund (2005) observes, is also a mountaineer learning to know his- or herself. Reflexive awareness and knowledge of place are therefore an "ongoing sensual dialogue between the surroundings and the self" (Lund, 2005, p. 29).

Beside touch, kinesis, and sight, smell and hearing also provide important sensory data. In a landmark study on the experience of ecotourism Waitt and Cook (2007) argue that ecotourists generate embodied knowledge based upon sensory experiences of their practices but also upon preconceived discursive notions of place and nature. The kayaker subjects of their study in fact felt the characteristics of the wilderness corporeally through their direct encounters with place, but also somehow managed to filter away and block stimuli and corporeal sensations that did not align with their preconceived notions of wilderness. For example, following the received meaning of wilderness as uninhabited place, one of their respondents removed from his perception of the riverine environment all traces of human presence, while others reinforced traditional boundaries between nature as something existing only "out there," far away from cities and villages.

Ecotourists' practice of adventure has caught researchers' attention also in relation to how easily wilderness can be commodified (something we explore in greater depth in Chapter Six) and how such commercialization of "wild" leisure is experienced at times similarly, and at times differently from the context of serious leisure. Stebbins (1992, p. 3) defines serious leisure as "the systematic pursuit of an amateur, hobbyist, or volunteer activity that participants find so substantial and interesting that, in the typical case, they launch themselves on a career centred on acquiring and expressing its special skills, knowledge, and experience." "Career" is not meant here in the traditional sense of paid employment, but more in the sense of a protracted and meaningful biographical trajectory and lifestyle commitment that end up informing a person's identity. In contrast to serious leisure (though the contrast is not always so clear; on this see

Kane & Zink, 2004), packaged adventure tourism practices mix risk, danger, and adrenaline with a great degree of routine, safety, structure, planning, control, and organizational oversight—all factors that contribute to the feeling of an insulated adventure (Schmidt, 1979). Nonetheless, even packaged adventure can be an important signifier of who one is and who one aspires to be (Kane & Tucker, 2004) because the skills, knowledge, preparation, sensation-seeking tendencies, goal-completion motivation, and experience (see Burke *et al.*, 2010) needed by packaged adventure tourists are not unimportant and, as Varley (2011) argues, often even more meaningful than traditionally defined risk for our understanding of adventure.

Furthermore, traditional serious leisure-style experiences of wilderness recreation and packaged wilderness adventure tourism have been found to be similar along the ritualistic dimension as well. Both experiences in fact provide participants with opportunities to escape mundane life—even if the responsibility for safety and risk management is assigned to professional guides—and therefore to experiment with alternative identities and ways of establishing solidarity with others. In Turner and Turner's (1978) words, wilderness adventures can be understood as *liminoid* activities: practices occurring at the margins of the central political and economic processes typical of individualized societies, and practices which cross over some imagined spatial and temporal borderline (Varley, 2011). Group-based liminoid activities in the wilderness have been known to provide participants with a powerful organic and radically democratic sense of collective immersion and belonging or *communitas* (see Sharpe, 2005; Varley, 2011).

Video 2 MUNRO BAGGING

Watch the video: https://vimeo.com/129221257*

Filmed in the spring of 2015 during a short trip to Scotland, this video documents the unique practice of Munro bagging through three interviews with British outdoor enthusiasts and through the perspective of three different outings. Munro bagging clearly exemplifies the

Continued

balance of order, control, wildness, risk, information management, and adventure typical of much contemporary outdoor recreation.

It is interesting to note that the filming segment from the Isle of Skye took place on land owned by the John Muir Trust of Scotland, an organization focused on the preservation of wild lands in Scotland. Munro bagging in general, in a sense, can properly be understood as a search for wilderness in a space where wildness is becoming very rare.

* Last accessed February 3, 2016.

WILDLIFE ENCOUNTERS

While daredevil explorations in the most remote corners of the world may captivate popular culture more than a mundane encounter with wildlife or a family's summertime nature-based tourism outing, it is through practices such as the latter that most people experience a sense of the wild. Nature-based tourism is a very broad category of travel inclusive of forms like ecotourism, agritourism, adventure travel, captive animal sightseeing, birdwatching, camping, and hunting and fishing. While the intensity of interaction with nature, the duration of the experience, costs, infrastructural set-up, and its organization and context may vary greatly, nature-based tourism depends on key constant factors such as the existence of relatively undisturbed natural areas (a very problematic concept, of course), the opportunity to acquire knowledge through interaction with natural environments, the practice of sustainability and conservation-driven principles, the possibility of economic gain, and of course its continued value as a form of leisure.

Nature-based tourists themselves may also be quite different from one another. Lindberg (1991), for example, identifies casual (the least committed), mainstream, dedicated, and hard-core nature tourists (the most invested in wanting to learn about nature and seeking its protection). In another, often-cited classification (Kusler, 1991), nature-based tourists are categorized into school and scientific group members (the most willing to "rough" it), the DIY independent travelers, and the ecotourists on tours (who demand organization, comfort, and safety). Regardless of classifications, several characteristics are common for the enjoyment of nature-based tourism in the wilderness, such as the opportunity to practice skill and apply knowledge (as seen in the section on adventure travel), the possibility to practice mental health and spiritual rejuvenation (see section on seeking wellness in wild places), and the chance to encounter wild animals, as we discuss here.

The continued, seemingly universal, and powerful appeal of wild animal encounters is arguably an outcome of the growing isolation of urbanized people from animals, accompanied by the sustained media glorification of fauna as the most appealing "Other" of human civilization (Curtin & Wilkes, 2005). In this broader sociocultural context, psychologists, sociobiologists, environmental psychologists, deep ecologists, and of course (more-than-) human geographers continuously reveal new aspects of the relationship between humans and wildlife, progressively uncovering the importance of interacting with wildlife for the sake of human well-being. Followers of the biophilia hypothesis (Wilson, 1984), for example, argue that humans have an innate emotional connection to other living organisms which has evolved over millennia to help us survive and adapt to our environments. Nonetheless, modern cultural values, urbanization, and industrialism have contributed to an increase in the sense of alienation of humans from animals and have simultaneously promoted a culture of domination and subjugation, which recently has taken the form of paternalistic protection (Vining, 2003). It is in light of this schizoid attitude that we can understand a great deal of the approach typical of nature-based tourism encounters with animals.

The most typical aspect of the experience of interaction with wildlife is wonder and awe (e.g. see Curtin, 2009). Nature-based tourists typically recount their encounters with wildlife in terms of excitement, amazement, mystery, delirium, and thrill, frequently commenting on the inability of words to express their complex feelings (Bulbeck, 2005; Cloke & Perkins, 2005; Curtin, 2009; Modelmog, 1998). Wild animal encounters, despite their brevity, can also be cathartic, inspirational, humbling, and can lead people to temporarily forget quotidian concerns, to feel self-realized, and to sense intimacy with the world (Bulbeck, 2005; Curtin, 2009). Animal encounters work as liminoid spaces and times, where people can contemplate life and pursue escape from modern existence (Price, 1999). Interactions with wildlife are also understood to be multi-dimensional embodied experiences through which people actively engage and provoke their senses (Abram, 1996; Van Hoven, 2011). Whether with grizzly bears (Van Hoven, 2011), alligators (Keul, 2013), dolphins (Curtin, 2006), whales (Cloke & Perkins, 2005), or polar bears (Lemelin & Wiersma, 2007), encounters with wild animals (or even those in semi-natural environments) "are able to inform people's sense of what it means to coexist and share space with other bodies" (Keul, 2013, p. 931).

Despite a growth in scholarly interest in sensory perceptions of wildlife other than those based on visual perception, the sense of sight remains dominant both as a research topic and as a phenomenological dimension of people's experience. Hill and colleagues (Hill *et al.*, 2014), for example, argue that most of the rainforest visitors they studied narrated their travel in terms of what they saw. One of their research participants, for instance, observed: "Just watching a trail of ants building towers is interesting . . . they evoke a childish sense of wonderment. . . . We are part of it and it's wonderful to see that these things exist and the wonder of it all stays with you" (p. 75). The act of seeing is also the origin of the all-too-important emotional tension that distinguishes wildlife encounters from the less thrilling interaction happening in captive environments like zoos. Seeing an animal in the wild is more of a triumphant "spotting" than a casual "glancing," and is therefore an open-ended and fortuitous experience that is as exciting as it can be scary and unsettling. It is in fact not uncommon for nature-based tourists to experience anxiety and distress at the prospect of seeing wildlife, or to experience fear together with a more serene sense of fascination (e.g. see Van Hoven (2011) with respect to bears, Keul (2013) for alligators, Curtin (2006) for sea animals, and Hill *et al.* (2014) for wild pigs).

Animal spotting also happens to be at the basis of the popularity of wildlife photography, combining the technical ability to take good photos with the personal skill of being able to find animals in the wild. One of the most often-employed concepts in this regard is John Urry's idea of the tourist gaze which, while not crafted for the purpose of understanding wildlife encounters, still manages to capture well their visual significance. Urry (1990, p. 1) writes:

> The tourism gaze is about consuming goods and services which are in some sense unnecessary. They are consumed because they supposedly generate pleasurable experiences which are different from those typically encountered in everyday life. And yet at least a part of

> that experience is to gaze upon or view a set of different scenes, of landscapes or townscapes which are out of the ordinary. When we "go away" we look at the environment with interest and curiosity. It speaks to us in ways we appreciate, or at least we anticipate that it will do so. In other words, we gaze at what we encounter. And this gaze is socially organised and systematised . . . even in the production of "unnecessary" pleasure there are in fact many professional experts who help to construct and develop our gaze as tourists.

Thus the tourist gaze has been used to understand how the shooting of photographs has in many cases replaced the shooting of animals with guns and rifles (Ryan, 2000; Sontag, 1979) while maintaining a focus on exhibiting and displaying animals—photographically—as a way of "capturing" the world, expressing the skillfulness, bravery, and (often) manliness of the shooter, and facilitating a scientific (but for this reason no less controlling in its spirit) apprehension of animal life (Ryan, 2000). Tourist photographs of wildlife are also understood as vehicles for the expression of the status and identity of the photographer (Desmond, 1999; Bulbeck, 2005; Curtin, 2006; Lemelin & Wiersma, 2007; Urry, 1990)—a self-aggrandizing project which many tourism promoters have strategically capitalized on in order to sell access to bigger, rarer, and more exotic animal trophies.

The "wildlife gaze" (Lemelin & Wiersma, 2007) is in fact at the center of the commodification of the animal encounter within the context of nature-based tourism. Wild animals are not always easy to spot, therefore making them camera-ready for the more impatient tourists has meant creating not only captive environments (e.g. see Anderson, 1995; Curtin, 2006) but also "semi-natural" spaces where various forms of enticement make it desirable for wild animals to pop out of the bush for amateur photographers to snap their pictures. In this regard Knight (2010), for instance, has discussed the popularity of semi-wild monkey parks in Japan, where the feeding of monkeys at known spots and regular intervals has made their appearances predictable for a mass of tourists sensitive to convenient access and short on available time. Enticing alligators by feeding them chicken has had similar results in American swamps (Keul, 2013). While such "McDonaldized" (Ritzer, 1993) practices do not always detract from the power of the experience, what is obvious is that wilderness environments are very much "produced" in connection with tourist practices and with the dominant culture which continues to view nature as exotic and Other (and therefore highly seductive) but also foreign, dangerous, uncomfortable, and inconvenient (and therefore in need of taming—as Rantala (2010) observes in relation to moose-spotting tours in Lapland). But ironically, as we will explain further in Chapter Five, the demand of the more hard-core and independent nature-based tourists for pure, authentic, unmediated, and unregulated encounters with wildlife—something that would seem more ideologically preferable—may be the less environmentally preferable option. As Markwell (2001, p. 55) nicely puts it: "in attempting to transcend the boundaries between the tourist and nature, and thus perhaps have a more 'authentic' experience, ecotourists may be unwittingly contributing to greater environmental damage than other tourists who are content to 'stay within the boundaries' provided."

Hunting and fishing

If we were to estimate the popularity of hunting and fishing from the size of the academic literature available about them, we would conclude that people do not hunt or fish much. Even more interestingly, human–wildlife interactions—again, judging from the size of the scholarly literature—happen about 100 times more frequently in the protected environs of nature-based tourist activities than in wilder circumstances marked by a clear predator–prey relationship. Humans nowadays, it would seem, shoot and capture wild animals only with digital cameras and no longer with guns and rods. We could not be more mistaken. The more plausible reasons for the limited research on this topic, according to Lovelock (2008b, p. 3), "may relate to the fact that hunting and shooting are not generally popular pastimes of the educated middle class, and furthermore, that as a field of research the topic falls between the uncomfortable (guns, firearms) and the unforgiveable (killing Bambi)."

Hunting and fishing are important sports, ecotourism practices, and culturally meaningful traditional activities, as well as very socio-economically relevant sustenance practices. Certainly not all hunting and fishing happen in wilderness areas—many officially protected wilderness areas might even strictly prohibit such activities—but a great deal of them do happen either in marine or freshwater environments or in large tracts of land where "big game trophy" can be shot and killed by anyone with a license (not to mention by poachers). Fishing, for example, is a pastime officially practiced by about 10% of UK residents, 16% of Americans, 19.5% of Australians, and 23% of Japanese (see Lovelock, 2008b). Hunting is not as popular, but figures are still higher than one might think: 2% of New Zealanders, 0.35% of Australians, and 6% of American adults hunt (Lovelock, 2008b). In Canada the numbers are even higher, with about one hunting license existing for every ten Canadians.

In the context of this chapter it matters little whether hunting and fishing are ecologically sensitive practices. In other words, we are less interested here in the ethics of consumptive wildlife practices than we are in the experience and practice of them, as well as where they occur. The "where" is an important issue for us for at least two reasons. First, hunting and fishing are big businesses and as a result more money is invested globally every year to attract hunters' and anglers' dollars, euros, and pounds. Eden and Barratt (2010) remark, for example, on the growing popularity of stocked ponds where anglers can drive in, park, and fish just feet away from their vehicles and access the bathroom facilities, convenience stories, and snack bars that these complexes feature. The same could of course be said for big business game reserves and ranches—Texas is home to many of these—where shooting fish (or anything else that moves) in a barrel has become an effective marketing scheme. Second, *where* hunting and fishing occur—and here we are talking less about "canned hunting" and more about the traditional variety—matters because encountering wildlife has immediate consequences for a broader ecology that directly affects both the experience and the practice of hunting and fishing. As Nustad (2011) has shown, the history of Europeans hunting big game in Africa has evolved in parallel with the history of colonization first, and later with the equally troubling development of population displacement caused by narrow-minded conservation policies. So, while hunting may still happen in African wilderness areas, *who* is hunting *what* and *how* are very much contested issues.

The geographies of hunting and fishing also matter because for the larger part contemporary geographers and colleagues in related disciplines and fields look at hunting and fishing as leisure-centered practices more than as subsistence activities. Thus we now know little about the experience of killing wildlife for food, especially by non-Western hunters and anglers. However, we know a good deal about how hunting is practiced as a pastime in a way that solidifies male camaraderie at the exclusion of women (Bye, 2003), about how hunters practice "wayfinding" in the wilderness (Hill, 2013), how hunting and fishing are in large part a direct cultural response to the increasing predictability of modern urban life (Franklin, 2008), how fishing and hunting, in all their various specializations, are deeply stratified by class and habitus (Franklin, 1999; Mordue, 2009), and even about how the practices of hunting and fishing have come to symbolize national identity and a country's relation with its wilderness (Washabaugh & Washabaugh, 2000, p. 73). New research on women's participation in consumptive outdoor recreation has become more available in recent years, especially activities that are typically associated with masculinity. Hunting is a practice that women are increasingly participating in, although little research has focused on the female experiences of hunting (Metcalf, Graefe, Trauntvein, & Burns, 2015). Metcalf et al. (2015) surveyed a random sample of 750 big game hunters in Oregon and through their findings they suggest that female hunters can be classified under four categories: less-engaged hunter, family-oriented hunter, nature-sport hunter, and all-round enthusiast.

Even though we know much more about the ethics, sustainability, economy, and politics of hunting and fishing than we know about their embodied experience and practice, recent research by scholars such as Adrian Franklin has begun to correct this tendency. Franklin highlights how the "underlying attraction of these activities is the possibility of a highly sensualized, intimate and exciting relation with the natural world" (2001, p. 58; also see 1999) that does not rely so much on an alleged and much-maligned primal urge to kill or satisfaction in capturing animals, but rather on the "embodied and aestheticized experiences of nature" (2001, p. 63). Such experiences are deeply sensuous, involving sight, hearing, touch, and smell (not to mention the sense of taste stimulated in the act of eating game or seafood) and requiring deep involvement, education of attention, and skill development (Franklin, 2001). Hunting and fishing also demand geographical awareness and the accumulation of local knowledge on landscape, climate, and seasonal cycles, and thus ultimately "knowledge about and love of place in terms of its particularities" (Franklin, 2001, p. 72)—the latter a theme which increasingly surfaces in hunters' self-defense of their sport in response to conservationist critiques. On the other hand at times the particularities of hunting and fishing, and thus the unpredictability of these activities, are greatly reduced by management schemes that permit hunting or fishing only when certain species are available in great numbers.

Box 4.3 DOING GENDER ON THE WEST COAST TRAIL

The West Coast Trail is one of North America's most internationally renowned hiking trails. The 75-km-long (47 miles) trail contours a remote section of British Columbia's southwestern Vancouver Island, crossing largely uninhabited forest and coastal regions. The trail—originally designed in order to provide search and rescue for naval accidents—has a moderate elevation gain but it is generally considered to be highly challenging because of its uneven terrain, steep ladders, typically wet weather, the need to negotiate tides, and the necessity to pack camping supplies (as no huts or similar facilities are available anywhere along the trail) and food for up to seven days of trekking. Park regulations limit entry to only 68 individuals per day, exacerbating a sense of remoteness and isolation.

I (Phillip) completed the West Coast Trail with three male friends in the spring of 2014. That experience gave me a chance to reflect on what it means to "do gender" on a wilderness trail. The notion of "doing gender" arises from recent theoretical developments in the field of gender studies which have come to treat gender not as an essence but rather as a performance. Gender, therefore, is continuously done by all of us, reflexively or unreflexively, in our habitual (and at times counterhabitual) mundane ways of life.

The idea of wilderness has regularly been criticized for its androcentric connotations. Critics argue that notions of wilderness and many of the practices unfolding therein regularly reveal ideologies and practices that pit wilderness travelers as explorers and conquerors of "virgin" land. The wilderness in turn works as a theatrical stage for the hyper-masculine will to display fortitude of character, test physical skill, engage in male camaraderie, handle risk, and master natural forces (see Brandth & Haugen 2005; Reed, 2003; Woodward, 2000). So, was this the case in our experience?

Trekking the West Coast Trail can be a relatively uniform affair, as it turns out, regardless of who undertakes it. You will wake up in the morning and get dressed in yesterday's clothes. You will make some breakfast for yourself, perhaps coffee over a freshly lit wood fire. You will walk a few steps to the outhouse, brush your teeth,

Continued

and use the flushless pit-style toilet. You will unpack your tent and roll up your sleeping bag and mat. You will walk. And walk. Ten, eleven, twelve or perhaps a few more miles—or maybe less, on a slower day. As you walk you will chat with your friends. And at times you will enjoy some quiet. At other times you will relish the chance to see, smell, touch, hear, and taste the West Coast. You will breathe crisp, fresh, marine air. And you will breathe hard at times, as you climb ladders, sink deeply into mud, struggle to gain traction in soft sand, and keep up the pace uphill. You will carry a heavy backpack—40 to 60 pounds depending on how much you thought you'd need on the trail. Then you will stop late in the afternoon. You will make fire. You will make food. You might go for a swim to clean yourself and to play. You'll talk and laugh with your friends. You will sleep better than you ever have, under the dark night's sky. Is all this a way of performing a hegemonic masculinity?

The answer depends on the boundaries we choose to draw around gender performances and what we consider appropriate conduct for sex categories. If we believe that gender scripts are strictly bounded by normative ideas that regulate who can act them, then we might view tests of physical fitness (e.g. walking, conquering fatigue) and outdoor survival (e.g. making fire, wayfinding) as occasions for the performance of traditional masculinity. Along the same lines we might think that women who engage in the same activities challenge stereotypical notions of femininity. But this approach might leave quite a few activities unaccounted for and a few explanations unwarranted. Talking, cooking (which is inevitably preceded by shopping and packing), gathering water, making and unmaking camp are not such masculine activities—in the traditional sense of the separate sphere idea. Given how much time these practices take up on any given day one might be tempted to say that trekking is as feminine (in the stereotypical sense of the word) as it is masculine.

But a better answer altogether might be to loosen up all the boundaries around gender performances and to dismiss binary-based notions. By dismissing traditional notions of the masculine and the feminine, and by embracing an idea of gender as a "style" of being and becoming that is unhinged from the masculine and the feminine and entirely unrestricted and open to experimentation, the idea of whether trekking in the wilderness is more or less masculine will suddenly lose purchase. A better (in our mind) approach to understanding styles of doing identity in the wilderness might very well be found in the idea of relational sensibility and the occasions for multiple, various, and gender-unbound performances that such sensibilities allow.

SEEKING WELLNESS IN WILD PLACES

A great deal of interdisciplinary research, often combining concepts and theories from education, psychology, and geography, has gone over the years into understanding the relation between the self and wilderness settings. Such research has at times unfolded as the direct consequence—as a type of program evaluation, as it were—of an increasing number of both commercial and not-for-profit initiatives centered around taking people into wilderness environments for a variety of mental, physical, and spiritual health purposes. The most common goals of these programs are therapy and rejuvenation.

Wilderness therapy

Wilderness therapy is a very popular form of social and psychological therapy in the US. About 1,000 such programs exist in America, with a growing number targeting specific ethnic-, gender-, and age-based populations. Simply defined, wilderness therapy is a type of treatment for a variety of clinical issues which unfolds in wilderness or wilderness-like settings. Typical clients include youths between 12 and 17 years of age—the majority of whom are reported to be white (98.4%) and male (65%) (see Rutko & Gillespie, 2013). Wilderness therapy is sometimes referred to as adventure therapy (Williams, 2000), wilderness challenge programs (Wilson & Lipsey, 2000), and outdoor behavioral health care (Russell, 2005). In their influential definition, Berman and Davis-Berman (1991) argue that wilderness therapy must essentially consist of the practice of taking troubled clients for several days into uninhabited areas where they strive to become self-sufficient under the guidance of competent outdoor education leaders and counselors.

Wilderness therapy is not a walk in the park. Russell (2001) finds that despite their variety in content and intensity, wilderness therapy programs must have a distinct therapeutic focus aimed at inducing personal change and self-growth by means of encouraging self-reflection, personal and interpersonal awareness, and skill cultivation. The most common clinical issues for which wilderness therapy is recommended include depression, family or school problems, low self-esteem, substance abuse, destructive disorders, and social adjustment disorders (Bettman *et al.*, 2013). An increasing number of families nowadays turn to wilderness therapy because it is perceived to be more successful for youths than traditional therapy and because it is less restrictive and stigmatizing than existing institutional alternatives (Bettman *et al.*, 2013). Among the most popular wilderness therapy programs are those offered by Outward Bound (Bettman *et al.*, 2013): a challenge-based program founded by German educator Kurt Hahn that is focused on experience-based and value-centered education targeted at facilitating skill acquisition and the development of character and maturity. In the US, wilderness therapy programs are state licensed and undergo periodic assessments to ensure safety. Many also offer after-care plans, formal evaluations, and professional counselor-led individual and group therapy.

There are two basic kinds of wilderness therapy: camp-based programs (a time during which participants remain in one location) and expedition-based ones. Both types, nonetheless, are characterized by physically challenging activities and by clients and staff living without modern conveniences such as light, plumbing, telephones, electronic technologies, or the shelter afforded by permanent building structures. For some kinds of programs staff intensively teach clients wilderness-survival and outdoor skills, whereas other programs are less education-centered and more concerned instead with giving participants plenty of time to think either in groups or alone as part of one-day-long "solo" experiences aimed at facilitating self-reflection. Family members may at times be present, but for most programs they are not. The length of programs ranges from a week to three months; however, the majority are less than two weeks in duration.

Wilderness therapy programs are based on creating conditions of duress and struggle for their participants. The pressures that clients experience during such challenging times elicit sentiments of stress, anxiety, and vulnerability, which in turn create opportunities for adaptive change and for recognizing the need for responsibly tackling social and personal issues. Program schedules are typically structured to gradually challenge participants to master more difficult tasks, thus gaining more and more self-esteem and confidence as well as a sense of resilience and a stronger personal identity (Russell, 2001). The caring, mentoring, and non-confrontational approach followed by the counselors is also meant to provide clients with a sense of trust and confidence in both others and self.

A good deal of research has gone toward discovering which factors wilderness therapy participants find more useful for making the most of their experience. Studies have highlighted in particular the importance of relationships with staff, the value of solo time, the challenges inherent in mastering new skills or enduring fatigue, the pleasure of being in natural surroundings, the intimacy of small group relationships, the sentiments of pride and accomplishment, the acquisition of a sense of courage, the feeling of being a valued team member, the sense of independence, and the time to reflect and be away from routines and schedules (see Bettman *et al.*, 2013 for a review). To give a vivid sense of all this, we believe it is interesting to reproduce the words of a participant—someone who happens to be an academic who had enrolled in a women's abuse survivor program:

> It is difficult to articulate the feelings that I have regarding this experience in my life. I am filled with a level of gratitude that is overwhelming. This was a life-changing experience for me, in ways that I am still becoming aware of. I have received several postcards that fill me with a sense of love and connectedness. I have also received my journal, which I can only look at in small bits of time, because it evokes such an intense emotional response. In trying to capture the essence of what this experience has meant for me, I am struck by the impact it has had on my spiritual development. I believe that the intensity of this result occurred for two primary reasons. I have always loved the outdoors and I have always used this love of the natural world as a means of solace and healing. However, I had never before spent 7 days in the wilderness. This aspect of this experience required me to completely disconnect from the chaos of the world; there were

> no cell phones, no computers, no electricity, no running water, or other modern conveniences. This level of immersion into the natural world allowed me to connect with my own sense of spirituality at a level that rarely exists for us in our busy lives. In addition, the intensity of my connections to the other group members bolstered my profound belief in the worth of every human being.
>
> (Kelly, 2011, p. 110)

Though a thorough evaluation of therapy programs is beyond the scope of our book, we should note that research shows that such programs are often, though not always, effective. Though the outcomes are not always clearly measured (or even particularly easy to measure to begin with) and though more data are available about the positive aspects of the experience in the immediate aftermath than in the long run, empirical evidence demonstrates that wilderness therapy programs are known to reduce both suicidal tendencies and recidivism rates of young offenders, improve psychological and social functioning, assist adolescents with behavioral problems such as being withdrawn and quiet, and are greatly effective at promoting significant growth in self-esteem and self-efficacy (Rutko & Gillespie, 2013). Evidence also points to how wilderness therapy programs can promote perseverance, strength, self-control, assertiveness, decision making, determination, extroversion, feelings of accomplishment, and leadership skills, and even challenge notions of gender roles and identity (Norton & Watt, 2013; Whittington, 2006). From a critical geography perspective, however, we should note that most of the research is rather uninformed or even outright uncaring about the spatialization of wilderness. In most studies, in fact, the wilderness is treated as a mere backdrop for interpersonal and personal dynamics, with little or no concern about how wilderness therapy programs construct a place as wilderness (Rutko & Gillespie, 2013).

Spirituality and self-discovery

Wilderness therapy programs are popular formal and structured opportunities for personal change that make use of the "rejuvenating" power of wilderness experiences; however, the restorative potential of wilderness outings does not necessarily have to be actualized through such structured programs and through the counseling and guidance they offer.

Self-realization, self-discovery, and the cultivation of spirituality in the wilderness may take place through informally organized individual or group outings or even through institutionally organized (corporate or non-profit) programs that combine some overhead planning with spontaneous interaction. An example of the latter is orientation wilderness trips offered by universities, for the most part American, to incoming first year students.

Collegiate wilderness orientation programs have been offered by American universities for three-quarters of the last decade. The first one was offered in 1935 by Dartmouth as a way to recruit new members for its Outing Club. Since then, well over 200 colleges have done the same to introduce about 17,000 students yearly to student-led leisure clubs in order to integrate them into campus culture (Bell, Holmes, & Williams, 2010). Research, in fact, shows a positive correlation between completion of wilderness orientation courses and retention (Murtaugh, Burns, & Schuster, 1999). Such positive outcomes are likely due to the fact that objective-based wilderness orientation programs help in facilitating high school graduates' transition to university life by providing them with self-esteem and confidence, a sense of responsibility and trust in others, various degrees of social capital, and problem-solving skills which they can later transfer to their new lifestyle and identity as college students (Bell *et al.*, 2008; also see Gass, Garvey, & Sugerman, 2003; O'Keefe, 1989). For example, in their study of a week-long backpacking-, kayaking-, and rock-climbing-based outing Lien and Goldenberg (2012) found that participants learned outdoor skills, "leave no trace" principles, and team skills which made the program very personally significant to each participant and to the group as a whole.

Some college wilderness orientation programs feature explicitly religious and spiritual characteristics in order to cultivate students' values, moral character, beliefs in the afterlife, and in order to nourish the "whole person", by imparting not just technical knowledge but also lessons in spiritual education (see Bobilya *et al.*, 2011). Seeking a sense of transcendence in the great outdoors—either as part of spiritually focused programs or outside of such initiatives—is no recent invention, of course. As Hitzhusen (2004, p. 41) explains, experiences of, and in, nature lead to feelings of awe, personal balance, inner peace, beauty, goodness, inspiration, renewal, intuition of the divine, ineffability, mystery, and a sense of the sublime. These typical characteristics of the spiritual experience—which James (1902) identified as ineffability (an inability to precisely describe the experience), noesis (the spontaneous acquisition of knowledge), transiency (the temporariness of the phenomenon), and passivity (a sense of lacking control over the experience)—are common across encounters with wilderness and inspirational enough to produce long-lasting effects. In their study Bobilya and colleagues (2011), for example, found that participation in the Wilderness Journey for First Year Students (WJFYS) program at North Carolina's Montreat College led to a stronger sense of community, an increased sense of competence, greater propensity to care about others and the environment, increased trust in God, and greater awareness of one's faith. The significance of such trips does not lessen, and indeed may even increase, over time due to the uniqueness of the experience, its timing in participants' lives, and the significance of its challenge in personal memory (Daniel, 2007).

Research on the spiritual dimensions of wilderness experience looks at both traditionally defined religious beliefs and non-religious spiritual experience. In this sense spirituality is more broadly understood as "a way of being and experiencing that comes about through awareness of a transcendent dimension and that is characterized by certain identifiable values in regard to self, others, nature, life, and whatever one considers to be Ultimate" (Elkins *et al.*, 1988, p. 10). An abundance of published studies tell us that there is no doubt that wilderness, and more broadly natural environments, are prime sites for the cultivation of spirituality (for a review see Heintzman, 2009). Amongst other findings, wilderness experiences have been found to be conducive to spiritual well-being (Heintzman, 2000, 2007a, 2007b), a trigger for spiritual discovery (Grafanaki *et al.*, 2005; Schmidt & Little, 2007), a way to connect with God's creation (Livengood, 2009), an opportunity for reflection on spiritual values (Brayley & Fox, 1998; Trainor & Norgaard, 1999) and for cultivating introspection (Heintzman, 2002), spiritual inspiration (Fredrickson & Anderson, 1999), and solitude (Fox, 1997; Riley & Hendee, 1999; White & Hendee, 2000).

There are several ways in which wilderness experiences facilitate spirituality. According to McDonald (1989) outdoor recreation allows people to feel simultaneously relaxed and challenged by a sense of adventure, sentiments which in turn increase sensitivity to one's surroundings, self, and others, as well as a heightened disposition to find meaning in things. Others (e.g. Fox, 1999) argue that wilderness outings may be initially characterized by fear and anxiety but subsequently—after the initial hesitations are won over—by a renewed sense of confidence, relaxation, tranquility, control, and comfort that allow for broader transcendental connections to be established between the self and the lifeworld. Traditionally religious people, according to Heintzman (2007b), also build connections between individual experiences and God's creation in this way. In an influential theoretical development, Kaplan (1995) has articulated these spiritual processes in what has become known as restorative environments theory.

The precise characteristics of the environments experienced and the practices that people engage in also matter a great deal. For example, Fredrickson and Anderson (1999) have argued for the importance of being in what they call "bona fide" wilderness where contact with nature (as opposed to, in the traditional binary opposition, traces of civilization) leads to deeper contemplative and reflexive experiences (also see Riley & Hendee, 1999; White & Hendee, 2000). Williams and Harvey (2001) have made similar remarks about the transcendent dimensions of traveling through forest environments (a practice occasionally referred to as forest-bathing). Being in so-called bona fide wilderness allows for a sense of being away from it all to emerge, and consequently for feeling a diffused lack of restraints, and a remove from routines, daily busyness, and traditional responsibilities (Heintzman, 2007b; Stringer & McAvoy, 1992). As for the precise activities undertaken, research has shown that a spiritual experience may take place during canoeing (Fredrickson & Anderson, 1999; Heintzman, 2007a; Stringer & McAvoy 1992; Swatton & Potter, 1998), backcountry adventures (Marsh, 2008), mountain hiking (Behan *et al.*, 2001; Stringer & McAvoy, 1992), environmental education course attendance (Heintzman, 2007b; Lasenby, 2003), observing and photographing nature (Heintzman, 2002), moments of being alone in the wilderness (Fox, 1997; Heintzman, 2007a, 2007b; Sweatman & Heintzman, 2004), camping (Heintzman, 1998; Sweatman & Heintzman, 2004), white-water kayaking (Sanford, 2007), surfing (Anderson, 2013), and even desert four-wheel driving (Narayan & MacBeth, 2009).

SUMMARY OF KEY POINTS

- Romanticism—intended as an ideological and cultural movement focused on the valorization of emotion, nature, originality, individuality, and passion—has provided us with the philosophical foundations for the exploration, contemplation, and conservation of nature.

- Wilderness recreation consists of "voluntary non-work activity that is organized for the attainment of personal and social benefit including restoration and social cohesion" (Kelly, 1996, p. 27) and that takes place in areas that participants consider to be wilderness or wilderness-like.
- The benefits of wilderness recreation practices have been found to include self-realization, awareness and respect for the natural environment, effective communication and leadership, creativity and inspiration, rejuvenation, spiritual development, and greater abstract and bodily knowledge of the lifeworld.
- Nature-based tourism is a very broad category of travel, inclusive of forms like ecotourism, agritourism, adventure travel, captive animal sightseeing, birdwatching, camping, and hunting and fishing. While the intensity of interaction with nature, the duration of the experience, costs, infrastructural set-up, and its organization and context may vary greatly, nature-based tourism depends on key constant factors such as the existence of relatively undisturbed natural areas, the opportunity to acquire knowledge through interaction with natural environments, the practice of sustainability and conservation-driven principles, the possibility of economic gain, and its value as a form of leisure.
- The most typical aspect of the experience of interaction with wildlife is wonder and awe. Nature-based tourists typically recount their encounters with wildlife in terms of excitement, amazement, mystery, delirium, and thrill, frequently commenting on the inability of words to express their complex feeling.
- Wilderness therapy programs are based on creating conditions of duress and struggle for their participants. The pressures that clients experience during such challenging times elicit sentiments of stress, anxiety, and vulnerability, which in turn create opportunities for adaptive change and for recognizing the need for responsibly tackling social and personal issues.

DISCUSSION QUESTIONS

1. In what ways have Romantic ideals informed your understanding of wilderness? Can you identify any particular artists and authors who have influenced your experience of wilderness?

2. Have you ever taken part in an organized form of travel such as a nature-based tour? In what ways was the experience different from an unstructured outing of friends or family?
3. Do you have any hunters or anglers as friends or family members? Why do they enjoy what they do? How are hunters different from anglers in their experiences of nature and wilderness?
4. Compare your experiences of encountering a large animal in the wild with your experience of encountering one in a confined, controlled environment like a nature park or zoo. How are the experiences different?
5. Are certain doses of risk and danger important to you for your enjoyment of the wilderness? Or do you prefer a controlled and safe environment in which risk and danger are managed by guides and experts? Can our wilderness experiences be the same without a degree of risk?
6. Why is adventure such a powerful experience for some people?
7. Have you ever spent time in the wilderness to rejuvenate yourself? What was your experience like?
8. Many people find wilderness sublime. But what does sublime mean to you? Have you ever felt a sense of the sublime in the wilderness?

KEY READINGS

Curtin, S. (2009). Wildlife tourism: The intangible, psychological benefits of human–wildlife encounters. *Current Issues in Tourism*, 12, 451–474.

Franklin, A. (2001). Neo-Darwinian leisures, the body, and nature: Hunting and angling in modernity. *Body & Society*, 7, 57–67.

Heintzman, P. (2009). Nature-based recreation and spirituality: A complex relationship. *Leisure Sciences*, 32, 72–89.

Ingold, T. (2000). *The perception of the environment*. London: Routledge.

Kahn, P. & P. Hasbach (eds) (2013). *Rediscovery of the wild*. Cambridge, MA: MIT Press.

Lovelock, B. (2008a). *Tourism and the consumption of wildlife: Hunting, shooting, and sport fishing*. New York: Routledge.

Michael, M. (2000). These boots are made for walking: Mundane technology, the body and human–environmental relations. *Body & Society*, 6, 126–145.

Mullins, P. (2014). A socio-environmental case for skill in outdoor adventure. *Journal of Experiential Education*, 37, 129–144.

Rutko, E. & J. Gillespie (2013). Where's the wilderness in wilderness therapy? *Journal of Experiential Education*, 36, 218–235.

Waitt G. & L. Cook (2007). Leaving nothing but ripples on the water: Performing ecotourism natures. *Social & Cultural Geography*, 8, 535–550.

WEBSITES

Society of Outdoor Recreation Professionals: www.recpro.org
The Association of Outdoor Recreation and Education: www.aore.org
Sierra Club: www.sierraclub.org
National Outdoors Leadership School: www.nols.edu
World Organization of the Scout Movement: http://scout.org/
Outward Bound: www.outwardbound.org

All of these websites were last accessed on December 2, 2015.

NOTE

1 For a critique of this "sportification" of natural places, and a set of solutions, see Wattchow & Brown (2011).

5

CONSERVING AND MANAGING WILDERNESS

Trails, Norway (Photo: April Vannini)

Wilderness conservation could seem like a benign endeavor. At first glance few goals would seem easier to achieve than setting aside a parcel of land to let it take care of itself. But in actuality few social and environmental objectives are riddled by more paradoxes, ironies, conflicts, and quandaries than wilderness protection (Turner, 1996). This is because many social agents have key stakes in wild lands. From conservation scientists, environmental managers, and recreation users to resource industries, indigenous groups, hunters, and tourist operators, wilderness areas attract scores of differently minded individuals, interest groups, state and non-governmental organizations, and local communities each with its own agenda, and all this contributes to making protected areas highly contentious. Therefore, leaving these places alone to take care of themselves is far from an ideal solution. In this chapter we will examine in detail the practice and politics of wilderness conservation.

We begin with a brief look at the law and policy of preservation by outlining how wild places are set aside and managed in select countries. Next we delve deeper into the modern evolution of wilderness conservation management. Subsequently, we analyze the problems and conflicts surrounding wilderness preservation—focusing in particular on issues of political ecology and indigenous land rights. The chapter then shifts to an extended reflection on how wild places are currently managed so as to achieve the dual purpose of environmental and social sustainability. Throughout the chapter we draw lessons from cases from many different countries around the world and from many different types of wild places, with some of them officially recognized as wilderness areas, and some *ad hoc* wilderness.

WILDERNESS LAW AND POLICY

Before we examine the political ecology of wilderness protection, let us begin by discussing how wilderness is officially set aside and managed through legislation and policy. Whereas throughout the rest of the book we have steadfastly maintained that wilderness goes well beyond the formal designation of an area as such, in this section we wish to limit our attention to official designations of an area as wilderness through laws and policies that protect it in accordance with Category 1b of the International Union for Conservation of Nature (IUCN) (see Chapter One, Box 1.1 for more information on these categories). While the rest of our discussion in this chapter will not focus exclusively on Category 1b protected areas, a brief look at "official wilderness" will set the context for our review of the conservation literature.

The primary objective of Category 1b is:

> to protect the long-term ecological integrity of natural areas that are undisturbed by significant human activity, free of modern infrastructure and where natural forces and processes predominate, so that current and future generations have the opportunity to experience such areas.
> (International Union for Conservation of Nature, 2014)

Additional objectives are:

- To provide for public access at levels and of a type which will maintain the wilderness qualities of the area for present and future generations;
- To enable indigenous communities to maintain their traditional wilderness-based lifestyle and customs, living at low density and using the available resources in ways compatible with the conservation objectives;
- To protect the relevant cultural and spiritual values and non-material benefits to indigenous or non-indigenous populations, such as solitude, respect for sacred sites, respect for ancestors etc.;
- To allow for low-impact minimally invasive educational and scientific research activities, when such activities cannot be conducted outside the wilderness area.

(International Union for Conservation of Nature, 2014)

According to the IUCN, these areas should:

- Be free of modern infrastructure, development and industrial extractive activity, including but not limited to roads, pipelines, power lines, cellphone towers, oil and gas platforms, offshore liquefied natural gas terminals, other permanent structures, mining, hydropower development, oil and gas extraction, agriculture including intensive livestock grazing, commercial fishing, low-flying aircraft etc., preferably with highly restricted or no motorized access.
- Be characterized by a high degree of intactness: containing a large percentage of the original extent of the ecosystem, complete or near-complete native faunal and floral assemblages, retaining intact predator–prey systems, and including large mammals.

- Be of sufficient size to protect biodiversity; to maintain ecological processes and ecosystem services; to maintain ecological refugia; to buffer against the impacts of climate change; and to maintain evolutionary processes.
- Offer outstanding opportunities for solitude, enjoyed once the area has been reached, by simple, quiet and non-intrusive means of travel . . .
- Be free of inappropriate or excessive human use or presence, which will decrease wilderness values and ultimately prevent an area from meeting [our] biological and cultural criteria . . . . However, human presence should not be the determining factor in deciding whether to establish a category 1b area. The key objectives are biological intactness and the absence of permanent infrastructure, extractive industries, agriculture, motorized use, and other indicators of modern or lasting technology.

(International Union for Conservation of Nature, 2014)

In addition wilderness areas can include:

> Somewhat disturbed areas that are capable of restoration to a wilderness state, and smaller areas that might be expanded or could play an important role in a larger wilderness protection strategy as part of a system of protected areas that includes wilderness, if the management objectives for those somewhat disturbed or smaller areas are otherwise consistent with the objectives set out above.
>
> (International Union for Conservation of Nature, 2014)

We must keep in mind that the IUCN is not a legislative or governing body; it can essentially only make recommendations. So how are wilderness laws and policies actually enacted and enforced across the world? In what follows we limit our attention to Australia, Canada, New Zealand, the US, and Iceland; countries with some of the most highly developed legal and administrative statutes and, as it happens, also countries which presumably house the majority of our readers (it should be noted that while there are many *ad hoc* wilderness places in the UK, especially in Scotland, there are no wilderness areas coherent with IUCN Category 1b there) (for more in-depth information see Carver, Evans, & Fritz, 2002). Readers interested in additional international cases are invited to consult the highly useful works by Kormos (2008) and Dawson and Hendee (2008).

We begin Down Under. Australia as a whole has one of the world's most thorough systems for identifying and protecting wilderness areas, a system that is currently characterized by a highly detailed management framework and informed by an improving spirit of public participation, community consultation, and even by shared ownership and co-management with indigenous communities—all this in spite of a highly contentious colonial legacy. Because of its large territory and its relatively concentrated population in coastal areas Australia is also in the position to continuously set aside new large reserves, whose management principles lately have notably shifted away from species-specific conservation approaches and toward more comprehensive ecosystem-based strategies. Yet, wilderness protection in Australia exists mostly at the state level and not so much at the federal level. Moreover, despite the high number of parks and wildlife reserves throughout all six states and two territories, only New South Wales and South Australia have specific purpose wilderness legislation. Important differences among specific management models also exist at the state level, making wilderness protection practices rather uneven from state to state. In spite of this, protected areas amount to 16.5% of the country's land mass—one of the largest ratios anywhere in the world.

Following Australia (which was second to the US) Canada was the third country in the world to create a national park. Though the formation of Banff National Park in 1885 had more to do with exploiting the tourist potential of mountain scenery and natural hot springs, the early onset of park legislation eventually allowed Canada to develop a vast and advanced system of protected areas, which now amounts to roughly 8–10% of the nation's size. With the exclusion of Antarctica, Canada is believed to be the home of 20% of the world's remaining wilderness. Though Canada lacks a federal wilderness law comparable to the US Wilderness Act (although as of the year 2000 federal wilderness areas can be protected under the amended Parks Act), wilderness protection exists at both the national and provincial/territorial level. Ontario, for example, has had a Wilderness Areas Act since 1959. This Act allows for the setting aside of land for the protection of flora and fauna, education, and recreation. Federally, the national park system outlined in Canada's National Parks Act is mostly managed by Parks Canada. A total of 42 parks and reserves protect 68 million acres. Key management principles include a zoning system within parks, the maintenance of ecological integrity, the prohibition of resource extraction, and consultation with First Nations (for a unique case, see Box 5.1).

Box 5.1: GWAII HAANAS NATIONAL PARK RESERVE AND HAIDA HERITAGE SITE

Since 1974, Moresby Island (Gwaii Haanas) of Haida Gwaii—a large and vastly undeveloped archipelago in British Columbia, Canada, 640 km north of Vancouver and 130 km west of the mainland—has been an intense clash site for competing ideas and policies of land use, ecological conservation, and aboriginal ownership.

In the 1970s, aboriginal traditional rights, land titles, and sovereignty became growing national and international concerns (Takeda & Røpke, 2010). At the time the provincial and federal government had only established treaties with First Nations for 3% of provincial Crown land. The lack of formal treaties signed with First Nations in BC and elsewhere in Canada went against Great Britain's Royal Proclamation of 1764 which stated that treaties must be signed before First Nations could be removed or disturbed from their land (Takeda & Røpke, 2010). In particular, the lack of land treaties and the unwillingness of governments to establish working relationships with First Nations resulted in dramatic conflict over land rights regarding extraction by logging companies. As Takeda and Røpke (2010) point out, "Since the 1940s, these 'crown' lands have been managed through a forest tenure system which established a set of property rights giving logging companies the ability to rent long-term harvesting rights for relatively little cost in exchange for maintaining basic processing operations" (p. 180).

Gwaii Haanas is the home of old growth forests, ecological diversity, and majestic landscapes, and it is of course of fundamental cultural importance to the Haida people. As a result, the logging dispute quickly gained popularity with non-indigenous environmentalists, islanders, and other British Columbians who jointly began to work together with the Haida Nation to fight for the protection of the land. "Concurrently," write Takeda and Røpke (2010, p. 180), "the Haida established the Council of the Haida Nation, a unified political body of the Haida people, and began to advance their comprehensive land claim with the federal government." In 1985, after unsuccessful bids to both the provincial and federal Supreme Courts to gain control of its lands, the Haida Nation designated South Moresby area as a Haida Heritage Site under the sovereignty of its hereditary chiefs. In response, in 1987 the Canadian government ceded its management responsibility in order for the South Morseby area to be designated as a national park (Takeda & Røpke, 2010).

Although the national park status meant that the area was protected from resource extraction, the sovereignty of the Haida people over the designated parkland and its management was still up for dispute. At the time the park was first established there was a distinct conflict concerning perspectives of land management between the Canadian government and the Haida Nation. National park status protected lands for tourists, whereas the Haida sought to reframe the area as a Haida Heritage Site as they envisioned South Moresby to exist for the continuation of Haida culture. This prompted the Haida people to propose a joint stewardship (Takeda & Røpke, 2010). In 1993 the Council of the Haida Nation and Canadian federal government reached an agreement that the designated area would remain under a co-management system, thus establishing the park as a reserve named Gwaii Hanaas National Park Reserve and Heritage Site. Unlike other national parks in Canada the 147,000 acre park is now explicitly focused on ecological and cultural protection and co-managed by the First Nation and the federal government.

Although co-management constituted a happy end to a long and intense dispute, the reserve is not without its contradictions and ongoing conflicts. Between 2002 and 2004 the government began the process of deregulating forestry which included making amendments to the Forest Act and Forest Practice Codes. As Takeda & Røpke (2010) observe:

Continued

> [T]he changes gave forestry companies an unprecedented degree of authority, while legislating the Province's own authority out. For example, measures to protect the environment could only be applied if they did 'not unduly reduce the supply of timber' and could not affect a forestry company's ability to remain 'vigorous, efficient and world competitive'.
> (p. 184)

The Haida Nation, environmentalists, islanders, and other organizations have had to continue to remain vigilant in protecting this pristine area of forest and wilderness through community meetings, consultations, court appeals, and direct action.

Websites

Coastal Stewardship Network
    http://coastalguardianwatchmen.ca/haida-gwaii-watchmen-program
Gwaii Haanas National Park Reserve and Haida Heritage Site
    www.pc.gc.ca/eng/pn-np/bc/gwaiihaanas/natcul.aspx
Council of the Haida Nation
    www.haidanation.ca

All of these websites were last accessed on December 2, 2015.

Similar to Australia and Canada, in New Zealand wilderness lies right at the core of national identity. There, three laws specify the definition, purpose, and management of wilderness: the Conservation Act of 1987, the Reserves Act of 1977, and the National Parks Act of 1980. Public conservation lands currently make up approximately 30% of the country. Wilderness areas are sometimes officially zoned within these broader lands, and at times only exist as *ad hoc* wilderness within the most remote and inaccessible regions of these reserves. Areas may be defined as wilderness when they are large enough to take two days to traverse on foot; have clear boundaries buffered from human influence; feature no form of development; contain ecological diversity; and have potential for recreation. In these areas New Zealand laws prohibit the construction of buildings and machinery; the use of motorized vehicles, aircraft, or vessels; the taking of animals; and the building of roads and trails. Scientific studies, limited and low-impact commercial activities, the managed introduction of species, search and rescue, and firefighting are all allowed activities, but they are carefully planned and controlled. Key challenges to the system over the years have included problems created by introduced animal species, regulation of aircraft access, and conflict and confusion over jurisdiction.

In the US, despite its controversial definition and embattled history the Wilderness Act has been very successful in protecting wilderness areas, at least as far as quantity is concerned. The US National Wilderness Preservation System (NWPS) is meant to protect wilderness areas at the federal level, though management is delegated to four different agencies: the National Park Service, the US Forest Service, the US Fish and Wildlife Service, and the Bureau of Land Management. Designation of land as wilderness occurs through an act of Congress, though some state and local authorities may also designate land in some cases. There are currently 758 designated wilderness areas, totaling 109,511,038 acres (even though this amounts to 4.8% of the land mass of the country, the number goes down to 2.7% if we exclude Hawaii and Alaska). Similar to other countries, American wilderness management aims to protect undeveloped habitats for both endangered species and human recreation. Key advantages of the US wilderness management model are its uniform and clear yet flexible standards, its regular reviews, the federal control of the system, and the extensive amount of public education that wilderness protection offers.

Though several countries in, or partly in, Europe (especially Finland and Russia) have well-developed wilderness legislation in place (and let us not forget that Greenland is home to Northeast Greenland National Park, which is the world's largest in IUCN Category 2 and an immense *ad hoc* wilderness), it is to Iceland that we wish to turn our attention as wilderness is of crucial relevance to Icelandic national polity and economy, both for its nature-based tourism industry and the threat of its hydroelectric development potential. Iceland's Nature Conservation Act of 1999 defines wilderness as an area of at least 25 km^2 where solitude and nature can be enjoyed without disturbance from human-made structures or motorized vehicles (these need to be at least 5 km away). The purpose of setting aside an area as wilderness, according to the same law, is to manage the interaction between humans and their environment in order to avoid harm to the air, sea, or water and to ensure that nature follows its own laws. Management strategies are in place to ensure that wilderness preservation facilitates access to knowledge of natural history and human heritage, as well as sustainable development. The law also allows for traditional subsistence activities. The country's main wilderness areas are located in the iconic Central Highland—a region featuring dramatic glaciers and volcanoes, lava fields, fast-flowing rivers, and geothermal forces. In total, Icelandic wilderness areas cover 9.9 million acres.

Extensive wilderness areas that are considered part of IUCN Category 1b exist in other countries too. In Europe these include, in order by size, Russia's Panaajärvi National Park (247,000 acres), Sweden's Fulufjället National Park (56,929 acres), Bulgaria's Central Balkan National Park (51,917 acres), Poland's Bieszczady National Park (45,510 acres), Bulgaria's Rila National Park (40,068 acres), Italy's Majella National Park (40,014 acres), Romania's Retezat National Park (35,111 acres), and Finland's Oulanka National Park (24,700 acres) (all these acreage figures refer to areas zoned as wilderness within these parks). Outside of Europe, North America, and Oceania, Category 1b wilderness areas are rare. In Africa the country with the most developed managerial and legislative frameworks for the protection of wilderness is South Africa, which is currently home to 741,000 acres of officially designated wilderness. Zimbabwe, Tanzania, and Namibia have somewhat less developed statutes but also sizable Category 1b wilderness areas. In Asia it is Japan, Sri Lanka, and the Philippines that have developed the most comprehensive administrative frameworks. Yet it is countries like China and India that are home to the largest regions of *ad hoc* wilderness, regardless of classification. In Latin America no Category 1b areas exist, despite the extensive network of protected areas with *ad hoc* wilderness regions within them.

MANAGING WILDERNESS

After an area has been officially designated for conservation an effective management framework must be created and enacted. This is done in order to ensure that the protected area is used as intended by law and protected from threats, illegal activities, and other undesirable events. The management of wilderness areas has to consistently abide by numerous regulations and professional best practices that inform the operation of a vast number of social agents: from high-level government officials all the way down to wardens (see Hermer, 2002; Lawson, 2003). While an exhaustive guide to wilderness management is beyond the scope of this book we want to outline a few guiding principles of contemporary wilderness management.

Wilderness management is a highly complex activity. Whenever a problem arises managers must quickly take into account numerous possible conflicting solutions, diverse bodies of specialized knowledge, and diverging sets of values and priorities, and cope with the limited monetary resources available to achieve their goals (Eagles & McCool, 2004). Though narrowly defined, widely accepted best practices are very rare in this highly differentiated context, the set of principles that Dawson and Hendee (2008, pp. 180–192) have recently amalgamated from the specialist literature contained in publications such as the *International Journal of Wilderness* and through knowledge networks such as the World Wilderness Congress go a long way toward guiding managers in solving daily problems and shaping long-range policies (also see Manning & Anderson, 2012). The 13 principles are as follows:

- manage wilderness as the most pristine extreme on the environmental modification spectrum;
- manage wilderness areas and wilderness sites within parks following a concept of non-degradation;
- manage human influences;
- manage wilderness biocentrically to produce human values and benefits;
- favor wilderness-dependent activities over activities that could take place elsewhere as well;
- guide wilderness management using written plans with specific area objectives;
- set carrying capacities as necessary to prevent unnatural change;
- focus management on threatened sites and damaging activities;
- apply only the minimum tools, regulations, or force to achieve objectives;
- monitor conditions and reflect on multiple opportunities to inform long-term stewardship;
- involve the public;
- manage wilderness in relation to adjacent lands;
- manage wilderness comprehensively, not as a system of parts.

The last three principles, in particular, focus on quite possibly the most fundamental idea behind contemporary wilderness management, the idea that wilderness is a vast and complex *ecosystem* and that it must be managed as such. Ecosystems are knots of linkages among all the various biotic and abiotic parts that comprise them. Every ecosystem can be understood in terms of its particular composition, its structure, and functions. It is therefore a grave mistake to think of wilderness as a natural system somehow separate from human society. Such separation would be an ecological impossibility; think of how something like human-caused climate change can affect even the most remote corner of Antarctic wilderness. From this point it follows that wilderness management is not an oxymoron; wilderness ecosystems necessarily comprise human agents, so people might as well be directly involved in managing or, better yet, "stewarding" them through their activities.

Ecosystems undergo constant change, but ecologists did not always believe this. For a long time throughout the twentieth century it was believed that ecosystems would eventually reach a climax: a stable "end" point that could last indefinitely unless the composition of an ecosystem was altered. In many ways this erroneous belief drove much of early nature conservation: it was simply thought that an ecosystem could be preserved indefinitely at this climax stage if its "pristine" conditions could last forever. But periodic disturbances—both natural and human-caused (e.g. fires, floods, climate changes, etc.)—affect ecosystems continuously over their lifetime (e.g. see Mann, 2005). Rather than focusing on finding the climax stage of an ecosystem and somehow "freezing" it there, wilderness managers therefore now focus on the specific nature of change by attempting to understand *historic ranges of variability* (HRV), or in other words, how much change specific ecosystems have endured and can endure. Knowing that human disturbances have affected and currently affect ecosystems' HRVs, the purpose of contemporary wilderness management is then to contain disturbance within acceptable levels, "repair" an ecosystem from unwanted change, and ultimately preserve the character of wilderness as specified by a particular policy or framework for an area (Eagles & McCool, 2004).

With all this in mind we should now be absolutely certain of the fact that wilderness is *not* defined by the absence of humans and therefore that direct involvement of human beings through wilderness management is inevitable and indeed even desirable. In fact the US Eastern Wilderness Areas Act (PL 93–622) challenged the original Wilderness Act definition in an effort to protect landscapes with considerable human engagement. This was a significant event that broadened how the US conceptualized wilderness and wilderness management. In fact, throughout the planet, humans are part of every ecosystem—though they are not always a necessary part—and early assumptions about the pristine character of areas such as certain North American places have been discounted by research that has shown that native populations hunted, burned, raised crops, and built settlements throughout areas later believed to have been utterly free of human presence (e.g. see Cronon, 1983; Denevan, 1992; Landres *et al.*, 1999; Mann, 2005; McCann 1999a, 1999b). Contemporary wilderness management, therefore, focuses not on keeping environments "pristine" but rather on mitigating the impact of modern technology and excessive human influence. Such mitigation may include, but it is not limited to, restrictions on animal grazing, prevention of loss of animal and plant species, control of introduced (non-native) plant and animal species, limitations on human access (these may be absolute or relative), restoration of primitive conditions, and fire management. *Achieving* wilderness in this way—a more logical goal than protecting it by simply drawing lines around it and leaving it alone—requires cycles of specialized knowledge-driven planning, execution, and monitoring.

If we believe that change is inevitable, then the process of wilderness protection begins with the determination of how much change is acceptable within a clearly defined area. For example, wilderness managers need to determine, among other indicators, the *carrying capacity* of a protected area, or in other words how many and which species it can sustainably be home to (Eagles & McCool, 2004). Studying the carrying capacity of a protected area also means deciding how many visitors should be allowed, which is a notably controversial process influenced by many scientific and political variables (Eagles & McCool, 2004). One of the most effective frameworks to determine and manage carrying capacity is known as *limits of acceptable change* (LAC). The LAC approach utilizes broad goals, precise objectives, key indicators, and widely accepted standards to evaluate the impact of visitors on a protected area's biophysical and social conditions. It is especially valuable in contentious areas where several conflicting goals clash (Eagles & McCool, 2004).

Dawson and Hendee (2008, pp. 224–233) identify ten steps for implementing the LAC approach:

1. Define goals and desired conditions;
2. Identify area issues, concerns, and threats;
3. Define and describe acceptable conditions;
4. Select indicators of resource and social conditions;
5. Inventory existing resource and social conditions;
6. Specify standards for resource and social indicators for each indicator;
7. Identify alternative opportunity class allocations;
8. Identify management actions for each alternative;
9. Evaluate and select a preferred alternative;
10. Implement actions and monitor conditions.

So, for example, you could begin by defining desired conditions in light of existing legislation, local conditions, and cultural values. Secondly, you would identify what characterizes a particular protected area. You might, for instance, realize that it provides a critical habitat for an endangered species but also that it is regularly used by a small number of local native hunters who follow traditional subsistence practices. Thirdly, you would determine an "opportunity class" for the wilderness area and the related acceptable conditions. An opportunity class "defines the resource, social, and managerial conditions considered desirable and appropriate within the wilderness" (Dawson & Hendee, 2008, p. 226). At that point, for steps 4, 5, and 6 you would need to select indicators of acceptable conditions, as well as desired standards, and continuously collect data about them over time. Then, steps 7, 8, and 9 are all about reflecting on your data. Through reflection you might determine what alternative opportunity classes could better fit your ecosystem's needs, or you might identify alternative approaches to maintain your limits of acceptable change. And finally, through ongoing monitoring you will be constantly alerted to current conditions and trends in relation to your preferred standards.

Contemporary and popular for wilderness management include strategies for resilience-based stewardship and collaborative adaptive management models (e.g. Armitage *et al.*, 2008; Chapin, Kofinas, & Folke, 2009) as well as practical tools used to monitor trends, improve on-the-ground stewardship, review policies, and implement change based on reliable data such as those contained in *Keeping it Wild* (Landres *et al.*, 2008), a strategy booklet aimed at coordinating the activities of the agencies involved in managing wilderness in the US (National Wilderness Preservation System). *Keeping it Wild* outlines strategies to select cost-effective measures focused on preserving wilderness character; collect data over time to assess trends in selected measures; and utilize trends to assess and report. This set of tools makes use of the qualities of wilderness identified in the 1964 US Wilderness Act, further subdividing each quality into monitoring questions, indicators (see Table 5.1), and measures which allow for measuring trends.

In conclusion, contemporary wilderness management increasingly attempts to go beyond the idea of naturalness (see Cole & Yung, 2010). Traditionally, wilderness had been meant to be "natural," in the sense of a place where social interventions are off-limits. But going beyond this simplistic approach to wilderness as naturalness means taking a proactive approach to fixing the mistakes of the past and avoiding new ones. Should we unproblematically let climate change, for example, alter environmental conditions in the name of letting nature take its course? Or should we decide to intervene in biological, social, and physical processes in order to achieve our conservation goals, even though this means actively assisting the recovery of ecosystems that have endured damage, degradation, and destruction (e.g. see Box 5.2 on rewilding)? Going beyond the principle of naturalness means engaging in interventions, interventions which may range from lighting controlled fires to reintroducing predators to a region, or from thinning forests in order to assist species migration to culling overabundant species. As Cole and Yung argue (2010, p. 7):

> Intervention implies exerting human control to compensate for human impact on the land. Because naturalness implies both a lack of human impact and a lack of human control, one of the meanings of naturalness will be violated whatever is done (or not done). Where interventions are pursued, decisions must be made about how to intervene, and well-supported management objectives and desired outcomes must be articulated. Objectives and outcomes must be knowable, attainable, and desirable. By most definitions, objectives based on naturalness have none of these attributes.

Table 5.1 The qualities, monitoring questions, and indicators for monitoring trends in wilderness character (Landres *et al.*, 2008, p. ii).

Wilderness character quality	Monitoring question	Indicators
Untrammeled—Wilderness is essentially unhindered and free from modern human control or manipulation	What are the trends in actions that control or manipulate the "earth and its community of life" inside wilderness?	Actions authorized by the federal land manager that manipulate the biophysical environment; actions not authorized by the federal land manager that manipulate the biophysical environment.
Natural—Wilderness ecological systems are substantially free from the effects of modern civilization	What are the trends in terrestrial, aquatic, and atmospheric natural resources inside wilderness?	Plant and animal species and communities; physical resources. Biophysical processes.
Undeveloped—Wilderness retains its primeval character and influence, and is essentially without permanent improvement or modern human occupation	What are the trends in non-recreational development inside wilderness?	Non-recreational structures, installations, and developments; inholdings.
	What are the trends in mechanization inside wilderness?	Use of motor vehicles, motorized equipment, or mechanical transport.
	What are the trends in cultural resources inside wilderness?	Loss of statutorily protected cultural resources.
Solitude or primitive and unconfined recreation—Wilderness provides outstanding opportunities for solitude or primitive and unconfined recreation	What are the trends in outstanding opportunities for solitude inside wilderness?	Remoteness from sights and sounds of people inside the wilderness; remoteness from occupied and modified areas outside the wilderness.
	What are the trends in outstanding opportunities for primitive and unconfined recreation inside wilderness?	Facilities that decrease self-reliant recreation; management restrictions on visitor behavior.

Box 5.2 REWILDING

The first definition of rewilding has been credited to John Muir's wilderness movement and conservation practices (Brown, McMorran, & Price, 2011). Rewilding gained popularity as a conservation theory in 1991 with the introduction to the Wildlands Project (now called the Wildlands Network) developed by conservationist Michael Soule and wilderness activist Dave Forman (Soule & Noss, 1998). *Wild Earth*, an environment magazine published in the US by the Wildlands Project, developed a special issue in which it shared the mission statement of the Wildlands Project and wilderness recovery strategy:

> The mission of the Wildlands Project is to help protect and restore the ecological richness and native biodiversity of North America through the establishment of a connected system of reserves. . . . It is time to . . . begin to allow nature to come out of hiding and to restore the links that will sustain both wilderness and the spirit of future human generations.
>
> (Foreman *et al.*, 1992, p. 3)

The Wildlands Project proposed a reserve model for wilderness recovery with the aim to "protect wild habitat, biodiversity, ecological integrity, ecological services, and evolutionary processes" (Foreman *et al.*, 1992, p. 4). However, the notion of a reserve model also came with a substantial modification to the existing nature-reserve model where large predators (keystone species) along with other species would also be protected. Wilderness is not determined by any particular land protection status, but rather "true wilderness" (Foreman *et al.*, 1992, p. 4) is a function of "wildness."

The term "rewilding" is enjoying a massive increase in popularity worldwide. "Rewilding" is used to describe a conservation practice that is distinguished from biodiversity conservation or habitat restoration—which can be defined as relating to the preservation of specific biotic or abiotic elements for the protection of certain

Continued

threatened species. Instead, rewilding aims to restore missing or dysfunctional ecosystems via a process of species reintroduction.

In North America, rewilding projects have seen a reintroduction of keystone animals in large designated corridors such as Yellowstone Park. Across Europe rewilding has become a growing practice. Within the UK there are several well-established rewilding projects. Aiming to restore forests, landscape functions, and to reintroduce missing species, these rewilding actions are sometimes seen as appropriate solutions in a broader process of restoration. Rewilding projects are increasingly deemed as necessary processes in many sites because of the perceived degraded nature of the landscape (see Rewilding Europe's website, www.rewildingeurope.com).

The ideas of connected areas, keystone species, and large areas of wilderness have evolved with the growing popularity of rewilding because rewilding aims to *restore* (as opposed to protect) functions within an environment so that ecosystems can function independently, or with minimal human intervention (Brown *et al.*, 2011; Soule & Noss, 1998). There is a growing consensus that biodiversity conservation must move away from managing loss and toward active restoration.

Supporters of rewilding believe that the removal of species such as wolves can have a devastating effect on an ecosystem, which results in unintended loss of other species (Crooks & Soule, 1999). Ecosystem functions can be regenerated by reintroducing extirpated keystone species that drive these processes (Brown *et al.*, 2011; Hintz, 2007; Soule & Noss, 1998). As some scholars have reported, the reintroduction of wolves to Yellowstone National Park in the US had a positive effect on the park's fauna, flora, and habitats due to the change in movement and population of elk (Brown, *et al.*, 2011; Ripple & Beschta, 2006). Rewilding is not explicitly aiming to restore ecosystems back to a time before human contact in a particular area. Jorgenson (2014) argues that Rewilding Europe makes a distinction between restoration and rewilding:

> Rewilding is really not about going back in time. It is instead about giving more room to wild, spontaneous nature to

> develop, in a modern society. Going back (to when?) is not a real alternative, it is just nostalgia. Rewilding is about moving forward, but letting nature itself decide much more and man decide much less (p. 5).

Once reintroduction of keystone species occurs, managing wilderness and wildlife becomes less of a human initiative since extirpated species drive active restoration naturally.

Websites

http://rewilding.org
www.wildlandsnetwork.org
www.rewildingeurope.com/

All of these websites were last accessed on December 2, 2015.

CONTEMPORARY POLITICAL ECOLOGIES OF WILDERNESS MANAGEMENT

Over the past 20 years geographies of environmental conservation have shed much light on the international politics of wilderness management. As a whole, these studies have shown us that creating and managing wilderness areas—whether official IUCN Category 1b or *ad hoc* wilderness—is a deeply contentious business marked by social, political, and cultural processes that are increasingly global in nature. Conservation is now supported (if not largely controlled) by powerful supernational bodies such as UNESCO, the UNDP (United Nations Development Programme, which includes the World Conservation Monitoring Centre), the World Bank, the IUCN (which includes the World Commission on Protected Areas), Conservation International, the Nature Conservancy, the World Wide Fund for Nature, and the World Resources Institute (Young *et al.*, 2001; Zimmerer, 2006). Many of these agencies then turn the day-to-day operation of their projects and programs over to International Environmental Non-Governmental Organizations (IENGOs), to which many states directly delegate the day-to-day management of conservation policies. Zimmerer (2006) argues that the worldwide expansion of conservation measures has been subject to powerful political and economic globalizing forces which have informed spatial–environmental dimensions (e.g. legal, political, and territorial) as well as the sociocultural dynamics that inform the measures, including discourses of science, ecotourism, nationalism, and environmentalism.

Geographers interested in examining the precise ways in which protection functions in such a global context often work within the theoretical tradition of *political ecology* (see Robbins, 2004). Their research differs greatly from the more practical and "managerial" approach (mostly discussed up to this point in this chapter) that characterizes the business of professional conservationists. Political ecologists—in geography, sociology, anthropology, and political science—traditionally employ a critical and polemical approach to understanding how protected areas affect the people who live near or within them, and to analyzing how conservation impacts communities' economies and ways of life (e.g. see Box 5.3). Chief among their interests are reflections on the relation between nature and society as enacted by various protected area designations and management frameworks; the creation and enforcement of boundary and border issues (both local and international) as a result of protected area creation; the distribution of power and resources among the stakeholders involved in area protection; and the occurrence of unintended consequences of conservation (see Zimmerer, 2006). In what follows we will examine current geographical (and cognate) political ecology research on conservation management by focusing first on the main problematic issues, and subsequently by highlighting possible ways forward or solutions.

Box 5.3 CONSERVATION REFUGEES

Conservation has historically resulted in the displacement of people and therefore has caused much hardship in these people's lives. The removal of native peoples in order to preserve biodiversity has had the unintended consequences of violating the human rights of many people, but especially of indigenous people, in the name of preservation and protection. But who is to blame? In his critique, Dowie (2009) specifically points the finger at big business conservation organizations he calls "BINGOs" (big, international, non-governmental organizations) such as Conservation International, the Nature Conservancy, the World Wide Fund for Nature, the Wildlife Conservation Society, and the African Wildlife Foundation. Dowie argues that BINGOs lack an understanding that the natural and social world are connected. In principle, biologically speaking, ecosystems can function normally without humans. However, in the contemporary world ecosystems are deeply interwoven with human activities from a cultural, political, and economic perspective. Removing humans from "nature" in order to preserve wild spaces disregards the various ways humans are intrinsic to the socio-natural functioning of ecosystems.

According to Dowie (2009), BINGOs have contributed to creating millions of conservation refugees throughout the globe. These organizations control large budgets and are also conduits of billions of dollars from bilateral and multilateral aid donors such as the World Bank, the European Union, and the United States Agency for International Development. Thus, because of their large wallets, they are able to determine conservation policies and practices that have a serious effect on the livelihoods of indigenous people. Many of these organizations support exclusionary conservation policies that infringe on indigenous rights by way of privileging ecosystem preservation. Because habitat preservation is perceived as the greater good, the removal of local peoples from their homelands is presented as an unavoidable necessity. Yet, Dowie (2009) argues, the consequences of exclusionary conservation practices are highly damaging to the purpose of conservation since people living within or nearby protected areas often become enemies of conservation—which they view as intruding on their everyday needs. Thus BINGOs' neglect of local knowledge and their ignorance of indigenous people's pre-existing stewardship of protected areas ultimately put strict conservation practices at risk of failing.

Website

www.conservationrefugees.org (last accessed December 2, 2015).

Wilderness management and its discontents

The single greatest cause of discontent in the way that protected areas are managed lies within the ideal of *human absence*. When we believe that nature and culture are separate we find it logical to exclude people from protected areas. Such exclusion has been known as the "fortress model of conservation": a model based on the idea that nature is best conserved inside the "walls" of clearly delineated and strictly regulated protected area boundaries (e.g. see Hermer, 2002; Roth, 2008). The fortress model has, however, been deeply criticized for dispossessing rural populations of their resources and homeland, for separating people from their land-based ways of life, and for denying communities the agency to manage their own territory. The splitting of the world into two purified ontological categories typical of this model—nature and science on one side and culture and politics on the other—is, according to philosopher of science Bruno Latour (2004), regrettably reinforced by a hierarchy of epistemologies that puts forth scientists and environmental "experts" as holders of the truth and politicians as partisan obstructionists (even when driven by genuine democratic ideals). So, even though more advanced community-based models of protected area management have been created throughout the world lately, fortress models of conservation remain widely practiced all over the globe and many strict environmentalists continue to strongly advocate against alternative models (Roth, 2008).

The fortress model dates back to the violent eviction of Native American populations from Yellowstone Park (see Burnham, 2000), a practice which has since been extensively copied throughout the world (King, 2010). The human absence ideal propagated by the Yellowstone approach has caused many managers to dismiss the importance of traditional practices followed by indigenous populations—practices which have subsequently been either forgotten or outright banned (Russell & Jambrecina, 2002). This has resulted in closures of indigenous camps and other habitations, restrictions on seasonal grazing, fishing, and hunting rights, and general disregard of indigenous culture and heritage. Indeed the very notion of wilderness, as intended to imply human absence, is now offensive to many aboriginal populations who more coherently view nature as deeply integrated with cultural life (Russell & Jambrecina, 2002). While over the past decade or so several countries have attempted to better recognize the importance of protected lands for indigenous peoples, new policies on cultural landscapes have not always been efficient or fair. For example some people can still be excluded from protected areas if they are viewed as not local or indigenous "enough" (Sundberg, 2006), or if they are members of a disadvantaged ethnic group or social category (e.g. women) whose land rights may not be evenly recognized (Holmes, 2014).

Many human geographers have argued that the construction of an abstract concept of space built for the absolute control of territory which is facilitated by centralized and distant management aims insidiously to secure national security and sovereignty more than sustain biodiversity (Roth, 2008). The human absence ideal and the way it is managed in order to regulate what protected areas mean, what purposes they should serve, whom they should benefit, who gets to decide how they work, and why, when, how, and who should access them has therefore been at the center of innumerable conflicts—conflicts that some scholars have interpreted through the lens of territorialization (see Holmes, 2014). Territorialization, explains Holmes (2014, p. 1), involves "delineating a particular space, determining what behaviour and activities are and are not allowed within it, giving it particular political and social meaning, and communicating this delineation and meaning to others." Territorialization "is a political process, serving particular social, political, or economic ends, pursued to make control over space easier" (Holmes, 2014, p. 1).

Violent evictions resulting from the territorialization of protected areas have been subject to a great deal of humanitarian outcry over the first two decades of the new millennium (e.g. see Dowie, 2009; Neumann, 1998; Spence, 1999; West *et al.*, 2006). Not only are protected area management models based on human exclusion and environmental isolation unproven to work (e.g. see Stevens, 1997) but it is clear that populations "decoupled" from their environment (Hoole & Barkes, 2009) inevitably end up suffering from long-term social, cultural, and economic disruption—sufferings dangerously exacerbated by variables such as ethnicity, class, caste, and gender (Craig *et al.*, 2012; Holmes & Brockington, 2013). In some cases displacement is further troubled by what happens after evicted populations reach their new territories (McLean & Straede, 2003). Lunstrum (2015), for example, details how the land initially set aside to accommodate the 7,000 people displaced by the formation of southern Africa's Great Limpopo Transfrontier Park has since been given to private investors seeking space for a 37,000 hectare sugar cane/ethanol plantation. What is particularly insidious is that such "green grabs" as these forms of "land grabs" are sometimes known, are viewed favorably by some ecotourism operators and internationally funded ecosystem project managers interested in climate change mitigation, thus revealing the extent to which private commercial interests, government agendas, and global political forces have occasionally become allies in the neo-liberal, neo-colonial economy.

Neo-liberalism, let us recall, is a system founded on economic principles of deregulation, privatization, and liberalization and political ideals such as individual freedom, democracy, and entrepreneurship. Neo-liberals all over the world lobby for "communities" and "civil society" to take on responsibilities traditionally reserved for the state. By delegating power to NGOs, voluntary institutions, and private organizations, states can divest themselves of not only duties but also—advocates of neo-liberalism argue—financial burdens. This is all for the better, neo-liberals suggest, as state agencies are inefficient, irresponsive to local needs, and slow, whereas local groups and businesses (often in partnership with each other) can better impact social and economic life.

Just prior to the time of writing, a small number of nation states have begun to abandon the fortress model and the human absence ideal, and made strides instead toward recognizing indigenous land claims. In Australia, for example, co-management schemes have sprung up all over the country between states and aboriginal groups. But while these accords are motivated by good intentions, their functioning in practice is not always so respectful of indigenous subsistence rights. For example, Palmer's (2004a, 2004b) research shows how fishing in Kakadu National Park has been subject to a great deal of controversy and conflict which has pitted Aboriginal (Bininj) and non-Aboriginal anglers against one another. Even though 50% of the park is Aboriginal land, and even though the park lease agreement stipulates that the whole land must be managed as *ad hoc* Aboriginal land, non-Aboriginal "territorians" have made claims that fishing is an essential expression of the territory's lifestyle and have therefore made demands for equal fishing access. The democratization of environmental protection has been as uneasy elsewhere. In Guatemala, for example, Sundberg (2003) reports how the implementation of the Maya Biosphere Reserve legislation has excluded local inhabitants from decision making in numerous ways.

Restrictions on subsistence rights are especially jarring in light of the injurious colonial history of wilderness preservation in the developing world. In Africa, for instance, the history of wildlife protection cannot be separated from Westerners' thirst for big game hunting (see King, 2010 for a review). Anderson and Grove (1987; see also MacKenzie, 1988, 1990) have shown how big game hunting played a crucial role for early African explorers in their attempt to establish themselves within the high ranks of the colonial society. And yet later on, after the colonial impulse turned its attention toward wildlife reserve formation (see Neumann, 1998), restrictions on local populations' hunting practices have resulted in cattle market collapse (Brockington, 2002; Hodgson, 2001), loss of traditional ways of life (Shetler, 2006), and hunger (Neumann, 1998). While all of this happens, sport hunting continues throughout Africa, often in combination with high-end ecotourism development. McDermott Hughes (2005, pp. 174–175) explains:

> Southern African conservationists imagine a continental space for tourists and investors and village spaces for peasants. At the continental scale, economically minded ecologists seek to heal the scars of partition and to let game and game viewers run free. They dream the African dream, and they dream of making money. Indeed, this form of planning treats future profits as certain—it is a conjuring act with respect to time. According to this line of thought, the cattle ranges of the Great Limpopo hold the potential for high-value tourism and sport hunting. These zones are optimum habitat, a desired third nature that is treated as real. Thus, Great Limpopo planners see the future while looking at the present. Their African scale affirms growth and denies boundaries. Meanwhile, the village scale denies growth and affirms boundaries. The "appraisers" of peasant societies describe a small-scale, bounded present while signs of an expanded, boundary-busting future are all around them. Above the din of children, planners speak of stable rural populations contained behind electric fences. But the fence, like the children, recedes from view. The Great Limpopo construes the village scale not as a limitation but as an opportunity for smallholders. Inequality—in the greatest feat of conjuring—becomes parity. After all, each party does according to its liking within its own space. Through intensive community-based projects, peasants can cultivate their gardens within their boundaries. In doing so, they free up land for extensive bioregions and travel routes. Tourists expand as peasants are enclosed. Large, vibrant (white) bioregions nestle against small, static (black) villages.

The true long-term obstacle for global conservation, as this African example evidences, is that the most undeveloped regions (and therefore those with the highest potential of being considered wild) are also often those with the highest degrees of human poverty. Contemporary park formation, therefore, cannot be abstracted from current neo-liberal discourses on "sustainable development" and how often these use wilderness protection as a Trojan horse for multinational business interests (see King, 2010 for a review).

Not all rhetoric on sustainable development is inauthentic, of course, but a great deal of it seems to perilously continue the Western imperialist legacy of the past. At the most fundamental cultural level the eurocentric idea of a civilization focused on the controlling and taming of a wild nature which is separate from the human sphere—often a discourse lying at the very foundation of all nature protection—is obviously an act of colonization when imposed upon peoples whose original worldview and the natural world are more deeply interdependent (Palmer, 2007; Rose, 1996; Suchet, 2002). "Being in the position of overlord," Suchet (2002, p. 147) observes, "allows humans to impose practices of intervention such as domination and management." Meanwhile, at the most practical and administrative level, much of the day-to-day wilderness management occurs by way of foreign involvement into local affairs, itself a form of imperialism (Young *et al.*, 2001). In some cases the designation of protected areas is the direct outcome of international pressure, for example through environmental conservation clauses tacked on to international lending programs or through agreements that forgive portions of a country's foreign debt in exchange for the formation of protected areas. In other cases, as Sundberg (2006) explains, IENGOs directly or indirectly fund and control protected areas' management, as a result of many states' ongoing devolution of social and environmental responsibilities. In Latin America, in particular, US-based NGOs like Conservation International and the Nature Conservancy have a great deal of neo-colonial power over conservation priorities and agendas. "Bolstered by their claims to technical expertise, impartiality, and goodwill" argues Sundberg (2006, p. 240), "NGOs naturalize their vision of human–land relations as correct, thereby producing truths that are then embedded in conservation policies."

Another common management problem lies in boundary disputes. Protected areas rely on clearly delineated boundaries, but all borders are abstractions. Also borders are not static: they regularly "leak" when people, animals, commodities, microbes, and other entities cross them.

These leaks inevitably "challenge the fixity of territorial identities and the hierarchies that structure access to resources" (Valdivia *et al.*, 2014, p. 686). In Galapagos Islands National Park for example, border crossings of wildlife into neighboring farmland and fishing grounds, or those of introduced species into parkland, continuously force reorganizations of conservation and alter the balance of power relations with local inhabitants. Similar to animals and plant species, fire cares little about borders as Sletto (2011) explains with a focus on Canaima National Park, Venezuela, and the differing fire management practices and goals inside and outside park boundaries. In this regard, Sletto (2011, p. 197) observes:

> The practice of conservation planning is still premised on rigid boundary-making intended to minimize the ambiguities, uncertainties and vicissitudes that characterize tropical ecologies, and to ultimately produce exclusionary, tightly ordered conservation spaces. Although work in conservation biology, landscape ecology, and related fields now increasingly emphasizes network-based landscape-level conservation management and critically considers the important roles of ecological boundaries and edge effects in shaping vegetation mosaics in heterogeneous landscapes, conservation management in tropical forests continues to favor rigid and often simplistic delineations between protected and unprotected spaces. This leaves little room to grapple with the fluidity, unpredictability and ambiguity that characterize human–environmental relations, particularly in landscapes undergoing rapid social and environmental change.

The most obvious example of troublesome boundary crossers is that of poachers. It is all too easy to think of poaching as evil, but the story changes when we realize that in many cases poachers are impoverished indigenous groups whose hunting practices (and related cultural rituals) have been recently denied by newly formed protected area boundaries. Recent research has shown the multiple dimensions of the humanitarian crisis surrounding poaching, for example by decrying the "shoot-on-sight" policy that some governments have issued against poachers (Neumann, 2004) and the outright "green militarization" of some protected areas such as South Africa's Kruger National Park, where more than 300 suspected poachers have been killed since 2008 (Lunstrum, 2014). Green militarization refers to the "use of military and paramilitary personnel, training, technologies, and partnerships in the pursuit of conservation efforts" against poaching (Lunstrum, 2014, p. 816). These kinds of practices have been reported in Guatemala, Colombia, Nepal, Indonesia, Botswana, Cameroon, Kenya, Tanzania, Zimbabwe, Congo, and the Central African Republic (see Lunstrum, 2014).

Another frequently observed problem that wilderness managers face is conflict among stakeholder groups. As more and more protected area managers move toward community-based management models (as we will see in more depth in the next section) the fact that communities are not as united and homogeneous as previously assumed becomes more obvious (e.g. see Wilshusen *et al.*, 2003). "Communities" may have little in common besides a territory as they may experience deep fractures along lines of religion, class, caste, ethnicity, and political and economic ambitions (Magome & Murombedzi, 2003). Following the 1980 IUCN World Conservation Strategy (International Union for Conservation of Nature, 1980), which compellingly argued that conservation depends upon the involvement of the communities it affects, many international environmental organizations have spoken in favor of the direct participation of citizens in wilderness management. Yet, over the past decade many researchers have found that some community stakeholders access more benefits from conservation than others, that different community groups relate to protected areas in radically different and sometimes unpredictable ways, and that these imbalances affect the effectiveness of area protection management (e.g. see Agrawal & Gibson, 2001; King, 2007; Palmer 2004a, 2004b, 2007).

Finally, all wilderness managers face the problem that every organization faces: poor leadership and organization. Examples abound in the literature. Haynes (2013) shows how the policy regulating Australia's Kakadu National Park is complex and confusing, leading the area's traditional Aboriginal owners and the state to share power in an inefficient and contested manner. Conflict over governance and struggles over legitimation—as well as over issues like development and growth—are also reported by Jamal and Eyre (2003) with regard to Canada's Banff National Park. While discussing India's Dhauladhar Wildlife Sanctuary, Fischer and Chhatre (2013) articulate another problem common to many wilderness management models: ineffective and irregular enforcement of regulations, which in their case study led to a protected area at first existing only on paper and then later—upon a sudden change in enforcement strength and frequency—to sudden conflict with local populations. Poor leadership, mixed with the eventual withdrawal of outside agencies responsible for assistance and oversight, is also lamented by Balint and Mashinya (2006) in relation to their study of a community-based conservation project in Zimbabwe. Many more "managerial" and logistical problems, and the social and political consequences these generate, could be listed and easily take up a book of their own.

In search of solutions and best practices

The most common and effective solution to several of the problems outlined above is *community-based natural resource management* (for different terms, but related features of this management model see Roe & Jack, 2001). Several scholars have argued that communities have inalienable rights to manage their own affairs and address their own concerns (e.g. see Bryant & Jarosz, 2004), especially in light of the fact that management by distant authorities has been known to lead to economic and political injustices as well as to environmental conservation failures (Bonta, 2005; Dowie, 2009; Goldman, 2003; Neumann, 1998; Robbins, 2004). The relative success behind community-based management is simply explained: if people (especially when economically disadvantaged) can draw some form of income and/or other benefit from a protected area, and can have some degree of control over their territory's natural resources, environmental conservation will be more successful (see Bonta, 2005; Hackel, 1999; Himley, 2009; Hoole & Barkes, 2009; Fischer & Chhatre, 2013; King, 2010). In order to foster engagement and involvement, community-based management models have tailored various kinds of partnerships between local community groups and protected area managers on the basis of local needs, as well as cooperatively identifying management objectives and solutions to problems, sharing revenues, involving community members in research projects and programs, developing fair and equitable processes to deal with conflicts, and monitoring community satisfaction (see Russell & Jambrecina, 2002). One of the best-known models for devolving authority to local communities is the CAMPFIRE scheme (Communal Area Management Programme For Indigenous Resources). CAMPFIRE's central principle is to aim for a cooperative wildlife management (see Balint & Mashinya, 2006; Marks, 2001; Wolmer, 2004). Briefly put, the CAMPFIRE program allows for multiple resource use (e.g. wildlife, livestock, and crops) in buffer-zone areas as a way of taking the strain off protected areas.

Despite their relative success, community-based models require improvement. For instance, in addition to the fractures existing within communities as discussed earlier, community-based models are subject to corruption, lack of accountability, and inequitable or limited resource distribution, and they sometimes find it difficult to deal with the challenges arising from shifting interests, as well as going beyond their democratic rhetoric in actual practice (e.g. see Balint & Mashinya, 2006; Himley, 2009; King, 2007, 2010). Blaikie (2006), for example, finds that at times community-based schemes mask unequal power relations under a veneer of democratic control and that external forces such as regional and national elites still continue to take the lion's share of the revenues from some of these programs. Conservation scholars, more so than political ecologists, are also skeptical toward community-based management as they believe that it does not always reduce threats to biodiversity (e.g. see Chapin, 2004; Terborgh *et al.*, 2002).

Often, community involvement in wilderness management leads to *mixed land use* agreements. Under these agreements a protected area is the site for various kinds of economically productive but sustainable activities (e.g. see Zimmerer, 2006; Zimmerer & Young, 1998). One of the best-known such management models is the biosphere reserve. Developed by UNESCO's Man (*sic*) and the Biosphere Programme in the 1980s to address environmental degradation simultaneously with socioeconomic development, the biosphere reserve model has been praised by several observers as an ideal alternative to fortress-style parks (see Sundberg, 1999). The model hinges on a process of zoning whereby an area is divided into a zone in which human residence and use is restricted, a multiple-use zone where people live and work, and a buffer zone between the two where restrictions on several kinds of activities are imposed. Other mixed-use models and arrangements exist, of course. Dempsey (2011) and Clapp (2004), for example, report on the "ecosystem-based management" protecting the two million hectares of British Columbia's Great Bear Rainforest—a unique (however, also very controversial) agreement among environmental groups, the provincial government, and the forest industry. Similar mixed-use innovative models have been adopted in China (Coggins, 2000), Holland (Arts *et al.*, 2012), and France (Alphandery & Fortier, 2007), only to name a few countries.

A key component of community-based conservation is the involvement of indigenous groups through the direct application of their traditional environmental knowledge (TEK) to conservation goals (Watson *et al.*, 2003). TEK is not based on a dualistic understanding of culture and nature and it directly promotes ecocentric respect for non-human entities and their recognition, appreciation of the importance of the meaning of local places, and recognition of ecological principles of interdependence (Watson *et al.*, 2003). TEK has great value for cooperative resource management because it allows wilderness managers to better understand the historical role of animals, plants, and people within an ecosystem (Hunn *et al.*, 2003; Watson *et al.*, 2003). Hunn and colleagues (Hunn *et al.*, 2003), for example, report how an appreciation of the TEK of the Huna Tlingit people of southeastern Alaska has been playing an instrumental role in understanding sustainable resource management. By allowing the Huna Tlingit people to sustainably harvest glaucous-winged gull eggs—as they historically were able to do on their land—Glacier Bay National Park and Preserve managers have been also successful in dramatically improving their relations with the surrounding native communities. A more mutually respectful and cooperative ecosystem-based management has also been reported in the Yukon's Kluane National Park, where TEK and Western science have found ways to merge (Danby *et al.*, 2003). While these management models are not beyond critique, within North America in particular there has been a significant trend toward the devolution of protected area management to indigenous groups and even toward the formation of tribal parks (see Carroll, 2014).

Another key stakeholder group that wilderness managers have learned to take into account is recreational visitors (Eagles & McCool, 2004). Wilderness experiences have been found to be shaped by three key factors: the *social* conditions experienced by visitors (such as the number of other visitors encountered), the *resource* conditions (such as the traces of human impact on the landscape), and the *management* conditions (such as the type and amount of regulations imposed) (Hendee, Stankey, & Lucas, 1990). Surveys and interviews of wilderness recreationists have found that users of wilderness areas prefer to have as few encounters as possible with other people, as well as a pristine environment, and freedom from control (e.g. see Dorwart *et al.*, 2009; Lawson & Manning, 2002). But of course these are but ideals, and trade-offs among priorities are generally necessary. Wilderness managers have therefore learned to take into account the preferences of their visitors in order to make decisions on specific trade-offs (Eagles & McCool, 2004; Lawson & Manning, 2002), which may, for example, include the acceptance of strict regulations and fees imposed on access (an obviously undesirable factor) in order to limit the number of daily visitors to an area, or the building and maintenance of clearly visible trails and boardwalks in order to prevent the soil erosion caused by wandering hikers.

Finally, a popular solution to the problem of "leaky borders," as discussed in the previous section, has been found in cross-border or transboundary protected areas and wildlife corridors. Transboundary protected areas are essentially multinational parks and reserves whose territories spread over two or more nation states. These areas are typically formed around "hot spots": zones of concentrated biodiversity rich with endemic species (King, 2010). Transboundary conservation unfolds along lines of biocentric and ecological interest rather than political borders, typically involving peace-promoting joint management (hence the nickname "peace parks" for these protected areas). The first peace park, Glacier National Park, dates back to 1932 when it was formed to join two protected areas across Montana and British Columbia. Shortly after the beginning of the new millennium, 169 transboundary projects in 113 countries existed (see King, 2010). Cross-border parks make sense ecologically and also geographically: typically it is peripheral areas lying at the edge of a country that remain underdeveloped (geopolitical borders are often drawn across mountain ranges, etc.). In some cases border zones are also left undeveloped as a way of providing a buffer between countries, such as the case of the former "iron curtain" between Western Europe and the Soviet Empire—now the subject of interesting biodiversity conservation initiatives (Schwartz, 2005; Zmelik *et al.*, 2011). An innovative type of cross-border project is wildlife "corridors": interlinked habitats that connect larger protected areas providing a means for wild animals and plant seeds to move between their refuges. Arguably the best known is the Y2Y, the Yellowstone to Yukon Conservation Initiative, which is working on an integrated conservation agenda and wildlife corridors for the immense region it encompasses.

SUMMARY OF KEY POINTS

- The International Union for Conservation of Nature (IUCN)'s protected area Category 1b is used by many countries worldwide to define and protect wilderness.

- Wilderness areas as defined by Category 1b are unevenly distributed around the world, as they are primarily concentrated in countries that have adopted official wilderness laws and policies.
- Wilderness is a vast, constantly changing, and complex *ecosystem* and it must be managed as such.
- Wilderness managers nowadays focus on the specific nature of ecosystem change by attempting to understand historic ranges of variability (HRV) and determining their limits of acceptable change (LAC).
- Contemporary wilderness management increasingly attempts to go beyond the idea of naturalness.
- Conservation is a profoundly political and global affair supported and often largely controlled by powerful super-national NGOs.
- Political ecologists have found that conservation is troubled by numerous problems, many of them originating in the "fortress" model.
- Community-based natural resource management models have made substantial progress in solving the problems generated by the human absence ideal.

DISCUSSION QUESTIONS

1. Would it be appropriate to let nature take care of itself in light of climate change?
2. Are IENGOs' interventions in wilderness conservation an act of colonization or are they justified in practice?
3. Is the IUCN definition of Category 1b appropriate?
4. Is your country's legislation on wilderness effective? What are its main problems?
5. Are there cases when the principle of going beyond naturalness does not work well?
6. In what ways can the principles behind carrying capacity become a problem?
7. Is the "fortress" model of conservation ever justifiable?
8. What problems besides those listed in this chapter trouble contemporary wilderness management?
9. What solutions besides those listed in this chapter can improve contemporary wilderness management practices?

KEY READINGS

Cole, D. N. & L. Yung (eds) (2010). *Beyond naturalness: Rethinking park and wilderness stewardship in an era of rapid change.* Washington, DC: Island Press.

Dawson, C. & J. Hendee (2008). *Wilderness management.* Fourth edition. New York: Fulcrum Press.

Dowie, M. (2009). *Conservation refugees: The hundred-year conflict between global conservation and native peoples.* Cambridge, MA: MIT Press.

Kormos, C. (ed.) (2008). *A handbook on international wilderness law and policy.* New York: Fulcrum Press.

Robbins, P. (2004). *Political ecology: A critical introduction.* Malden, MA: Blackwell.

Turner, J. (1996). *The abstract wild.* Tempe, AZ: University of Arizona Press.

WEBSITES

Yellowstone to Yukon Conservation Initiative: http://y2y.net/

International Journal of Wilderness: http://ijw.org/

The World Wilderness Congress: www.wild.org/main/world-wilderness-congress

Peace Parks Foundation: www.peaceparks.org

Society for Ecological Restoration: www.ser.org

Indigenous Peoples' Restoration Network: www.ser.org/iprn/traditional-ecological-knowledge

Conservation Corridor: http://conservationcorridor.org/

Journal of Political Ecology: http://jpe.library.arizona.edu/

All of these websites were last accessed on December 2, 2015.

6

UTILIZING AND EXPLOITING WILDERNESS

Three Gorges Dam, China (Photo: April Vannini)

Whether we like it or not, wilderness is a *resource*: there are no uncertain ways to say this. Though their diverse intents may be as different as subsistence, leisure, and conservation, and their stakeholders as varied as sustenance hunters, conservation biologists, nature tourists, and miners, from an anthropocentric perspective wilderness areas are nothing but resources of material and symbolic wealth—sometimes untapped, sometimes sustainably utilized, and sometimes utterly abused. Following such a perspective, in this chapter we will examine a multitude of ways in which wilderness resources are utilized and exploited by different individuals and groups with different purposes. By utilization we refer to the act of turning something to use. By exploitation we refer to the process whereby something is turned to practical account for the sake of economic gain. We do not employ the concept of exploitation in a Marxist sense, but we do believe that to exploit a resource means to take advantage of it, regardless of whether that use is fair, sustainable, or otherwise. In this chapter, however, we do not play the arbiters of what constitutes utilization and what constitutes exploitation. That is what environmental ethics treatises are for. We, instead, cover practices of utilization and exploitation alike: from sustainable ecotourism to the mining of resources in a protected area. While we realize that the consequences of these acts are dramatically different we leave it up to you to reflect on the gray areas that separate utilization from exploitation on a case-by-case basis.

If we view the environment as a resource, then we can understand it as providing different types of services to us. Following an ecosystem services perspective (Grunewald & Bastian, 2015), wilderness utilization and exploitation can be grouped under four categories. Under the first category, **provisioning**, wilderness areas may allow for the collection of food, water, and raw materials of multiple kinds. Under the second category, **regulating**, wilderness areas serve important roles in carbon sequestration and climate regulation. Under the third category of **supporting**, wilderness areas work as important resources for nutrient recycling. And finally under the fourth category, the **cultural** one, wilderness may be a resource for recreational experiences such as ecotourism, and for scientific and educational purposes. Our chapter focuses on the first and the fourth categories, provisioning and culture, and it is divided into two parts. The first, longer part focuses on non-consumptive forms of wilderness utilization and exploitation. By "non-consumptive" we mean acts that are based, in principle, on the idea of leaving wilderness resources more-or-less intact for the enjoyment of future users. These are practices that may appear on paper to be focused on utilizing, not exploiting. Non-consumptive practices include a variety of tourist activities such as ecotourism, wildlife tourism, and nature-based tourism (activities that are somewhat different from one another but, for our intents and purposes, relatively synonymous). After a general introduction to these forms of tourism, in order to glean in detail valuable lessons we zero in on a variety of empirical cases drawn from the Galapagos Islands, southern Africa, Iceland, and Australia. Subsequently, we turn our attention to consumptive forms of wilderness exploitation. In that latter part of the chapter we examine practices that appear to be exploitative, such as mining, fishing and hunting, logging, drilling, road building, and related activities characterized by the explicit intent of modifying wilderness areas and thus, potentially, extinguishing their wild character. But, as you will see, divisions between utilization and exploitation will not be so clear cut. By the end of the chapter it will be clear that the boundaries between non-consumptive and consumptive, provisioning and cultural services, and/or utilization and exploitation, are more blurred than we may think.

TOURISTS IN THE WILD

As discussed elsewhere in our book, wilderness conservation is far from a benign endeavor. Forcible evictions, limited or non-existent planning consultation processes with local residents, inequities in revenue distribution, poaching, resource degradation, and many other problems have historically beset conservation initiatives such as those inspired by the "fortress model" of conservation. A few decades ago, when the problems became too obvious to sweep under the carpet, park officials, planners, politicians, conservationists, and business groups worldwide began to envision alternatives to the original protectionist philosophy. The strongest emerging argument was that the success of conservation initiatives could be much greater if the people living within or nearby protected areas could financially benefit from environmental conservation. The case for ecotourism was thus born. The World Conservation Union (IUCN) defines ecotourism as:

> [E]nvironmentally responsible travel and visitation to relatively undisturbed natural areas, in order to enjoy and appreciate nature (and any accompanying cultural features—both past and present), that promotes conservation, has low visitor impact, and provides for beneficially active socio-economic involvement of local populations.
>
> (World Conservation Union, [n.d.])

Taking roots at first in Africa, ecotourism initiatives focused on putting wildlife reserves under the control of local councils which began receiving revenue from their wild lands and thus learning new ways to make low-impact, environmentally sensitive tourism sustainable ecologically and economically in the long run (Honey, 2008).

Given the success of the early experiments, ecotourism was set for a big boom. As neo-liberal ideologies spread like wildfire throughout the world in the 1980s, many 'Third World' countries began to see ecotourism money as a key earner of foreign currency: a source of long-lasting revenue much less destructive and often much more valuable than extractive activities. In Kenya, for example, the annual ecotourist value of a single live lion was estimated to be $7,000—while a live herd of elephants was valued as high as $610,000—and thus exponentially more than their value when dead. Similarly, the value of live fish in the Caribbean and whales in Iceland was found to be much higher as a magnet for diving, snorkeling, and cruise ship ecotourists than for consumption after capture (see Honey, 2008, p. 23). Armed with such knowledge, by the early 1990s many developing countries had made ecotourism a cornerstone of their economic strategy, with some of them, like Costa Rica (see Box 6.1), Kenya, Tanzania, Namibia, and Ecuador, actively branding themselves as ecotourism paradises. Fueled by international loan schemes, private–public partnerships, IENGO involvement, and multinational business alliances the idea behind ecotourism managed to tap into the growing international environmental consciousness and the increased delegation of public responsibilities such as nature protection to the private sector.

Box 6.1: COSTA RICA'S ECOTOURISM

Costa Rica provides an interesting and unique case study to explore the industry of ecotourism and sustainable development. Costa Rica's tourist industry has increased revenues by 28% since the 1990s with the introduction of ecotourism management models. Honey (2008) outlines that prior to the 1990s and the early development of ecotourism, Costa Rica had attracted mostly domestic and foreign visitors from Central America. But by the 1990s Costa Rica came to be seen worldwide as *the* ecotourism destination to visit, surpassing other destinations such as Kenya, Nepal, and the Galapagos Islands (2008, p. 160). Tourism soon after became the number one industry in Costa Rica.

Honey (2008) explains that the success and high profile of Costa Rica as an ecotourism destination can be attributed to a variety of factors. In Costa Rica ecotourism has been developed around its national park system, good governance, and functioning democracy. Compared with other developing countries, Costa Rica is politically stable, has saved money and gained in popularity by abolishing its army, adhering to human rights, being welcoming to visitors, boasting the highest standard of living quality and literacy in Latin America, providing its citizens with a public health system and public education, and being a leader in conservation science. Costa Rica also is the home to many international and local environmental NGOs, it is close to the US, and due to a relatively modern infrastructure (save for a few bumpy roads) it is very easy to get around.

Tourists in Costa Rica have also been reported to enjoy a variety of activities besides visits to national parks and wildlife viewing. These include beach and water activities such as snorkeling, surfing, and rafting, as well as hiking, tree canopy ziplines, and spa and yoga retreats. Honey (2008) states that "in 2007, the tourism ministry announced it would officially promote four types of tourism—eco-, adventure, sun/sand/beach, and rural community based" (p. 161). Although Costa Rica is often branded as an ecotourist destination, its government has developed a "two track policy" which supports both ecotourism pursuits and larger international hotels and tourists chains, thus creating a variety of choices of tourist activities and infrastructure. Such diversity has occasionally produced a negative impact on local ecotourism projects, policies, and planning, especially on the country's Pacific coast.

The guiding principles of ecotourism aim to ensure economic benefits and maintain the integrity of the natural environment. Despite the larger international hotel chains and beach resorts,

Continued

Honey (2008) does find that Costa Rica still provides a high level of ecotourism initiatives, especially considering that it is the second most visited tourist destination in Latin America, following Mexico. In spite of the high volume of tourists that Costa Rica receives each year, the country is still able to adhere to seven of the main criteria of successful ecotourism, which include minimizing impact on the natural environment, building environmental awareness, providing direct financial benefits for conservation, providing financial benefits and empowerment for local people, respecting local culture, and supporting human rights and democratic movements.

Website

Costa Rican National Chamber of Ecotourism
http://canaeco.org/en (last accessed December 2, 2015).

However, given its widespread acceptance ecotourism soon became the subject of a great deal of exploitative "greenwashing" as well. Greenwashing refers to promotional activities and discourses that attempt to pass something off as environmentally friendly—perhaps under the veneer of merely cosmetic gestures or perhaps simply through empty rhetoric—while in actuality unsustainable business continues as usual. As a result, the mere mention of ecotourism today is often met with skepticism, cynicism, and confusion. One of the key sources of unease about the ecotourism concept is the difficulty inherent in matching actual practice with the abstract ideals contained in its definition (Carrier & MacLeod, 2005). Carrier and MacLeod (2005) argue that these inconsistencies often generate an "ecotourist bubble." An ecotourist bubble, not unlike the better-known idea of a tourist bubble, refers to practices that are ecotourist on the surface but still result in divorcing tourists' experiences from the social, political, and environmental contexts of their practices. Aiming for a stricter, normative definition of tourism, Honey (2008, p. 208) posits that ecotourism should refer to:

> travel to fragile, pristine, and usually protected areas that strives to be low impact and (often) small scale. It helps educate the traveler,

> provides funds for conservation, directly benefits the economic development and political empowerment of local communities, and fosters respect for different cultures and for human rights.

While this definition is not without its problems, it works as an ideal benchmark. Nevertheless, as we will see, the reality is often different on the ground. To examine in detail some of the challenges of ecotourism we now focus on four cases meant to shed light on common themes and recurring issues present in the literature.

The Galapagos Islands

We begin in paradise (Salwen, 1989)—or so we would be led to believe by the popular image of the Galapagos Islands. Before the two of us traveled to the Galapagos in 2014 the remote archipelago felt farther to us than any other destination on earth. Indeed it seemed somehow insulated from the globe itself, almost as if inhabiting a social and natural microcosm of its own. And yet the Galapagos Islands, following Charles Darwin's fieldwork and arguably more so after the hundreds of nature documentaries dedicated to their unique endemic species with their odd adaptive features and their endearing lack of fear toward humans, could not possibly have appeared more *earthly* upon our arrival. A modern airport, taxis whisking travelers and locals to their accommodations and homes, and shuttle vans leading Bermuda-donning, camera-toting tourists to their cruise ships before they could catch a glimpse of the abundant litter on the streets, quickly burst our bubble. The Galapagos—declared by Ecuador "at risk" in 2007 and listed by UNESCO on its unenviable catalog of "World Heritage Sites in danger" in 2008—felt lovely, yes, but also deeply overrun by ordinary problems plaguing just about any ecotourist destination. Our ecotourist bubble had burst before we could even spot our first sea lion.

The first of those problems was, well, us: travelers. From 1990 to 2014 travelers to the Galapagos have increased four-fold, up to 170,000 a year—in spite of earliest restrictions that put a cap on a few thousand annual arrivals (Honey, 2008). Though the archipelago is believed to retain 95% of its original biodiversity, travelers—whether cruise ship tourists, conscientious land-based nature travelers, or scholars on research leave—inevitably bring modifications to local environmental conditions, ranging from accumulation of waste to overuse of fresh water, and from soil erosion along trails to animal behavior modification (Watkins & Cruz, 2007). All of this was of course largely unknown to our many friends and family members back home, who fantasized that we could only make our way to the islands by hitching a ride on an expedition vessel. Nothing could be farther from the truth (a similar reality gap revealed itself on the heavily trafficked summit of Mount Fuji (see Video 3)). Thirty flights per week arrive at and leave the Galapagos Islands' two main airports, conveniently connecting Quito's new international airport with dozens of global destinations. This should be little news to careful observers: from their very foundation as a national park the Galapagos were envisioned by Ecuadorean authorities as a magnet for nature-based tourism (Hennessy & McCleary, 2011; Larson, 2001).

A quick walk around the streets of Santa Cruz and San Cristóbal persuaded us that the many travelers did not fly all the way to the Galapagos for a Spartan getaway. With us we brought the need for restaurants, hotels, shops, tour companies, and all the ancillary services these businesses required to make our stay enjoyable. So, while you would be hard pressed to spot a single local human being on a Galapagos nature documentary, by spending a few nights in town you would quickly realize that there are a lot of them—30,000 year-round residents (Watkins & Cruz, 2007). Unfortunately, the problems which a growing population causes in the archipelago are serious. As the WWF reports:

> human population growth, invader species, and commercial fishing threaten to destroy the fragile ecological balance in the world famous Galapagos islands . . . Although 97% of the island's land area has National Park status, the population of the Galapagos islands has more than doubled in the past 10 years, mainly due to migration from the Ecuadorian mainland. With this migration, many foreign plant and animal species are being introduced. Their estimated numbers have grown from about 77 in 1971 to more than 260 today.
>
> (cited in Taylor *et al.*, 2003, p. 978)

Wallace (1993), the author of an influential study on the links between ecotourism and immigration in the archipelago, is equally pessimistic: ecotourism is failing in the Galapagos, he argues, because Ecuadorean officials have not been able to contain migration. By the beginning of the twenty-first century almost 70% of the Galapagos residents were migrants from the rest of Ecuador.

In spite of the heavy migration it is not clear whether *Gualapagueños* (the residents of the Galapagos Islands) are better off as a result of the growth in ecotourism. The reason why this is so became crystal clear to us after a few days on San Cristóbal: most travelers never set foot on the island, or if they did at all they were quickly shuttled away to the nearly uninhabited southeastern corner where they visited a semi-captive giant tortoise rehabilitation center. A similar phenomenon happened on Santa Cruz, where often cruise ship tourists would leave their floating hotels only to visit the Darwin Center, on the outskirts of Puerto Ayora. These practices meant that rather than patronize land-based businesses, tourists would spend their money and time in the Galapagos almost exclusively on one of the 90 ships—varying in size from 16 to 90 passengers—sailing along one of a handful of pre-determined routes mostly calling at outlying uninhabited islands. This not only short-changed the locals but also gave most travelers a highly partial impression of the archipelago as pristine, natural, and entirely free of humans (save for the scientists they might have run into at the Darwin Center). We had a first-hand confirmation of this when a friendly Chilean woman who had visited the Galapagos by cruise stared at the two of us in utter disbelief as we broke the news to her that there were people, 30,000 of them, who lived in the islands (also see Hennessy & McCleary, 2011).

Recent research on the impact of ecotourism in the Galapagos shows that not only the local population is largely hidden from tourist view but also the infrastructure and waste the ecotourism industry generates (Quiroga, 2009). Unsurprisingly, this results in inequalities in revenue distribution (Taylor *et al.*, 2003). Of the US$419 million dollars spent by tourists in Galapagos during 2007, only as little as US$62.9 million is estimated to have entered the local economy (see Epler, 2007). This is largely because airlines and large travel agencies or mainland- and internationally based tour companies book the majority of trips and take in most of the revenue—and these companies are not necessarily benefiting island-based operators (Epler, 2007; Quiroga, 2009). Even though this scenario is gradually changing due to the growth of Galapagos-based businesses and the increased market share of land-based travel, a large proportion of the revenue is still pocketed by a small number of wealthy island families (Epler, 2007). Inequities like these have led to strong conflicts among local groups, between local and government authorities, and also between locals and outside businesses and conservation stakeholders (Hennessy & McCleary, 2011). While the Galapagos were removed in 2010 from the "in danger" list by UNESCO's World Heritage Committee, and while new governmental policies have done much to improve the potential sustainability of tourist and migrant mobilities, IUCN and UNESCO remain cautious about the future of the islands.

Video 3 CROWDING ON THE SUMMIT

Watch the video: https://vimeo.com/101804153*

At 3,776 m (12,388 feet) Mount Fuji is Japan's tallest mountain. It's also its most venerated. But while popular pictorial depictions and poetic words frequently celebrate Mount Fuji as a majestic oasis of sublime serenity, modern-day Mount Fuji is more of a tourist mecca for time-challenged weekend hikers. Every year during July and August—the short season when the four trails to the top are officially open—no fewer than approximately 200,000 people attempt the ascent.

Mount Fuji's summit can be reached via four trails, one on the northern slope, one on the southern, one on the western, and one on the eastern. Each trail is divided into "stations," with the first station located at the very bottom of the mountain, and the ninth station being the closest to the summit. A station is an assemblage of different facilities. Fifth stations are by far the largest, as they are the places from whence most people depart. Typically found at the end of a paved road, fifth stations like the one at Kawaguchiko are home to bus terminals, parking lots, restaurants, cafeterias, and coffee shops, as well as convenience stores, souvenir and outdoor apparel stores, and accommodations and other tourist facilities. Lower and upper stations, normally only reached by foot, are miniature versions of fifth stations.

It normally takes about six to eight hours to summit Mount Fuji from the 2,305 m altitude of the Kawaguchiko Fifth Station. Stations six and seven both unevenly sprawl across the steep mountain slope in a terrace-like fashion, with closely clustered huts nearly overhanging each other. I (Phillip) stopped at both stations for a few minutes to recuperate from the visual exhaustion brought on by having to ensure that my stride wouldn't cause me to trip on the feet of the person ahead of me. Stopping for a few minutes was also a chance to talk with people. I had a good laugh when a California-born GI—together with colleagues on a weekend off from their Okinawa base—remarked to me that station seven felt just like Tokyo's Shinjuku Station. There was a lot of truth in that exaggeration—which I later tried to depict in this video.

* Last accessed February 3, 2016.

Iceland

Often referred to by marketers as "Europe's last wilderness," Iceland's geography is marked by a low population (c.329,000) heavily concentrated around the capital Reykjavik and by a massive area of wild lands known as the Highlands which constitute about 40% of the territory of the island nation. The uninhabited Highlands—as mentioned before in Chapter Three—feature extremely rugged, volcanic, glacial, and mountainous terrain punctuated by fjord coastal landscapes to the east and west, dramatic waterfalls, and wide-open vistas stretching across the deserted interior and its extensive lava fields. Phillip visited Iceland in 1994 and fell in love with both these landscapes and the very notion of wilderness they evoked. At that time Icelandic tourism was not as brisk a business as it is nowadays. While tourism to Iceland is an old phenomenon, tourist arrivals began to grow gradually from the 1980s—increasing from an average of 65,000 to ten times that figure in 2012 (see Olafsdottir, 2013b). To date, tourism is Iceland's second largest industry after fisheries and ahead of heavy industry (e.g. aluminum production), commanding a 23.5% share of the country's total export revenues.

There are many issues about wilderness exploitation in Iceland that we could highlight, but one in particular stands out as being unique and particularly revealing. It pertains to a nature-based tourist activity that Phillip himself engaged in, back in 1994: the Highland safari. Highland safaris are essentially RTV-led (Rough Terrain Vehicle) trips into the Highlands organized by local tour agencies to allow nature tourists easy access to glaciers, waterfalls, and lava fields. While these tours are no different in structure from any sightseeing coach tour, the uniqueness of RTV Highland safaris lies in the fact that these off-road vehicles are able to get just about anywhere. To put it in different terms, drivers of these vehicles—with all the necessary skills and caution—can carry a small number of passengers (in the case of RTV buses even up to 40!) deep into the wilderness with no care at all for the absence of roads. And to make matters even worse for the romantic amongst us who still believes that a place is not a wilderness unless it can only be traversed carrying but a "two weeks' pack" on one's back, tourists are far from being the only ones driving off-road vehicles in Iceland. As Huijbens and Benediktsson (2007, p. 151) report, in Iceland "a large and growing proportion of the general public now has access to vehicles with off-road capabilities, which are able to deal with the still mostly 'roadless' interior."

What is the problem with all this? Huijbens and Benediktsson (2007) explain that the trouble begins with conceiving nature as a "destination," an idea shaped by the social history of the automobile. "Without cars," Aronczyk (2005, p. 1) puts it in simple and elegant terms, "wilderness as we know it could not exist." Cars promise drivers and passengers unlimited access, freedom, effortless movement, and escape from the mundane drudgery of city life. RTVs and Highland safaris have therefore rendered the interior of Iceland a deeply ambivalent and contested space. On one hand these mobile activities carry great significance for both locals, as a source of a way of life and a national identity, and foreign tourists, as a way of getting back in touch with wilderness or perhaps even discovering it for the first time (as was the case for Phillip). On the other hand, as Huijbens and Benediktsson (2007, p. 162) argue:

> from the environmental point of view, the unsustainability of these recreational practices hardly needs elaboration, whether one looks at fuel consumption or effects on vegetation and soils. From the critical angle of cultural analysis, it is also all too easy to see these developments in a rather negative light . . . . They do involve a particular gendering; a technological fetishisation of a certain kind; and sometimes rather bold and brash promises by jeep importers, tour operators and others with commercial interests, of the possibilities the technology allegedly affords for "communing with nature."

It does not end there either. The growth of automobility in the Highlands has played a key role in fueling resource exploration. Iceland depends heavily on both hydroelectricity and geothermal power for residential and industrial energy consumption (the aluminum industry has a particularly voracious appetite for it). The Highlands' still largely untapped potential for power generation has prompted the Icelandic government to plan constructions that in spite of a negative environmental assessment will "cross 'untouched' lava fields, valleys, salmon rivers and fjords, transforming a landscape that is currently offered to and perceived by tourists as pure and wild" (Olafsdottir, 2013b, p. 133). Europe may end up not having its (alleged) "last wilderness" for too long.

Southern Africa

Due to factors such as the low level of industrial and agricultural development typical of the region, the size and number of conservation areas, the worldwide popularity of African wildlife such as lions, elephants, rhinos, the Cape buffalo, and the African leopard (i.e. the "Big Five"), the widespread presence of conservation-focused NGOs, and the need for economic development, the southern region of Africa is home to some of the world's most renowned ecotourist sites. In this context ecotourism has been sought as a way of promoting conservation, generating employment, providing revenue through park fees and tourist expenditures, and attracting investment capital for infrastructure development. Spurred by the popularity of initiatives such as payments for ecosystem services and integrated landscape programs (Barrett *et al.*, 2013; Egoh *et al.*, 2012; Milder *et al.*, 2014), and especially by Integrated Conservation and Development Programs (known as ICDPs), many southern African nations have worked hard at reconciling environmental conservation and socioeconomic improvement by forming public–private alliances of local and international stakeholders (Cagalanan, 2013; García-Amado *et al.*, 2013; Spenceley, 2008).

Despite some successes, ICDPs have their pernicious effects. While ecotourist initiatives may be much better from an environmental perspective than most consumptive activities are, their advantages are not as apparent as one may think. Southern African cases provide clear evidence of this. For starters, much of ecotourism in the region stands as a clear endorsement of Western consumerism and as a Trojan horse for capitalism to access small-scale, subsistence-based, traditional economies. Ecotourism—in spite of its ideals—may then result in the commodification of local cultures, in the displacement of resident populations from protected areas or from places highly sought after by tourists, in the exclusion of certain groups from decision-making processes affecting their livelihoods, and in casting local populations as pliable hosts charged with the duty of serving foreign paying guests (see West, Igoe, & Brockington, 2006).

Social, cultural, and environmental side-effects of ecotourism problems are often exacerbated by increasing tourist flows. The numbers, indeed, are staggering. Namibia, home to only about two million residents, sees one million tourist arrivals per year. Mozambique records twice as many annual arrivals, and South Africa is now approaching the ten million mark. High tourist flows like these do create jobs but the number of permanent jobs and skilled positions continues to be limited, therefore few opportunities for upward economic mobility exist (Mbaiwa, 2003; Okech, 2010). At the same time, with more and more opportunities and spaces for ecotourism more land becomes inaccessible to local populations for traditional activities and ways of life. Ecotourism, in other words, often ends up becoming an avatar for neo-liberalism. In light of all this, as West and Carrier (2004, p. 485) put it, ecotourism in developing countries necessarily involves:

> institutional structures and practices that seem likely to reshape [aspects of the social and natural world] to conform to the virtual reality defined by important Western models of society and nature. This reshaping underlies what we see as an important contradiction in ecotourism: its tendency to *lead* not to the preservation of valued ecosystems but to the creation of landscapes that conform to important Western idealizations of nature through a market-oriented nature politics that results in the creation of a product that fits the [ecotourist] market needs. There is also a tendency for ecotourism to lead not to the support of distinctive local sociocultural beliefs and practices, valued by ecotourists because they represent alternatives to capitalist market systems, but to the spread and strengthening of those systems.

Moreover, economic benefits may not be all they are cracked up to be. In a recent study comparing Namibia and Mozambique, Silva and

Khatiwada (2014) have found several factors that complicate assessments based on straight financial numbers. In fact, even though the households located in their ICDP study areas did have higher income than their control sites (76% higher in Namibia and 40% higher in Mozambique), interview data with research participants showed that locals did not perceive themselves as better off economically, especially when they did not have family members employed in the tourism business. Relatedly, their respondents were not more satisfied with life, citing, in particular, problems caused by both conservation areas and tourism. A 76-year-old Namibian woman explained: "The life I had before with my livestock is now nonexistent. We used to have livestock graze around the whole area without a problem. After the lodges and campsites were built, they limited our space for grazing" (Silva & Khatiwada, 2014, p. 37). Another respondent, a 55-year-old man from Mozambique, laconically observed: "I don't know what the government is thinking because they're the ones who say we should make our own living. But we have no land plots because they're being taken by CBNRM [community-based natural resource management]. My question is how are we going to survive or make this so-called living if they're taking our land?" (Silva & Khatiwada, 2014, p. 37) In sum, much work remains to be done in the region and elsewhere to decommodify ecotourism and make it more equitable.

Australia

For our last in-depth case we turn our attention to Australia. In particular, we zero in on the issue of heritage and indigenous rights—extremely important in the Australian context. And for that we begin at the Great Barrier Reef Marine Park, where April and I traveled in 2006. The Reef is an amazing place. It is the world's most extensive reef system, encompassing a fringe reef along the shore, the ribbon reef (found north of Cairns), and a platform reef (off the continental shelf). Together it covers 348,700 km^2 or about one-and-a-half times the size of the United Kingdom. It is home to fish, coral, and other forms of marine life which—blessed by a water temperature of 23 to 26 degrees Celsius—is a major attraction for scuba divers and snorkelers. Known the world over for its beauty, the Reef was granted World Heritage status in 1981 (today it is one of 17 such Australian locations). Most of its area is administered by the Great Barrier Reef Marine Park Authority (GBRMPA), which has divided the Reef into a variety of zones ranging from general use to strict preservation where entry is essentially prohibited. Tourism access, however, is allowed in 99.8% of the Reef (White & King, 2009).

Due to its popularity, the Reef is in danger of overexploitation. In 1991 the United Nations International Maritime Organization declared the Reef to be a "particularly sensitive" area. This was no surprise. Australian tourism authorities market the Reef widely and heavily. With the arrival of the first high-speed catamarans in the 1980s much of the Reef went from being accessible only to hardy groups of enthusiasts to reachable by masses of tourists who could suddenly enjoy it on a variety of day trips from multiple coastal locations. Tourism to the Reef nowadays, measured at around 1.5 million arrivals yearly, generates AU$5.8 billion annually and keeps 63,000 people employed (White & King, 2009). Most tourist activity is concentrated around Cairns and the Whitsunday Islands. Damage to the corals created by swimmers accidentally hitting or walking on the Reef is the most common environmental problem; however, nesting sea-bird, dugong, and humpback whale behavior has been impacted by the heavy tourist presence as well. On land, excessive and poorly planned development in the Whitsundays has been a highly contentious issue.

As White and King (2009) have rightly noted, the Reef has been mainly considered a kind of natural heritage rather than cultural or human heritage in spite of the long history of the rapport between it and Aborigines. This is an example of what Palmer (2006) has called a "culture of nature" which creates spaces that are naturalized and emptied of politics and culture (also see Head, 2000 in relation to a similar case in Australia's Summerland Peninsula). Today there are 12,000 Aborigines and Torres Strait Islanders living in Reef-adjacent cities and towns, and another 11,000 around the Cape York Peninsula, including 70 traditional "owner" groups on the coast from Bundaberg to the Torres Strait Islands. We need to remind ourselves that "ownership" does not assume the same meaning in this context as it would in a Western or European sense. Aborigines and Torres Strait Islanders view their rapport with land as one of kinship, not possession, meaning an inherited rapport of cohabitation. Nonetheless in 1992 the High Court of Australia issued the Mabo decision which recognized the inhabitants of Murray Island in the Torres Strait as owners, granting them native title. This effectively marked the end of the *terra nullius* concept (see Box 2.2) and opened the doors for granting title claims all along coastal Queensland. In spite of this, the social and economic realities of islands now reinhabited by indigenous peoples, such as Thursday Island and any of the island tourist destinations, are sharp.

Regrettably, the participation of traditional owner groups in the nature-based tourist industry of the Reef is extremely low, and so is their involvement with the GBRMPA with the partial exception of a very limited number of areas. To address the need for greater participation, in 1994 the GBRMPA announced a 25-year objective to allow Aborigines and Torres Strait Islanders to "pursue their own lifestyle and culture, and exercise responsibility for issues, areas of land and sea, and resources relevant to their heritage within the bounds of ecologically sustainable use" (Great Barrier Reef Marine Authority, 1994, p. 35). Three years later the Sea Forum group was created to represent their interests. Nevertheless, much work remains to be done to make indigenous groups the key players they should be in Australian ecotourism. This situation stands in sharp contrast to Kakadu (as discussed in Chapter Five) and Uluru National Parks. At Uluru—considered by several ancestral groups to be a sacred crossroads of dreaming tracks—joint management has resulted in Tjukurpa traditional law principles becoming infused into the protection of the park as a cultural landscape. This, for example, has meant that many tours are led by Anangu guides and interpreters who are keen on teaching tourists about their way of life and about the sacredness of the landscape. Anangu people have also played a key role in educating travelers about the inappropriateness of climbing Uluru (though the climb is still legal and thus perfectly exploitative of the place and traditional inhabitants' rights).

DRAWING LESSONS FROM ECOTOURISM

As the four cases described above have shown, ecotourist practices in wilderness sites are potentially beneficial to both conservation and economic development agendas, yet their potential often remains unachieved. Three categories of problems are routinely highlighted in the literature: environmental, social, and cultural. These categories of problems are inextricable from one another. Recent literature shows these problems exist across cases in each continent. Let us do a broad, final recap of the issues.

From an environmental perspective, the devil is all in the details of ecotourism. While in principle ecotourist practices are light and sustainable, in practice their impact can be massive, especially when tourist numbers go up. Spenceley (2005) lists the most common types of damage in each of the following categories (pp. 140–141):

- *Atmosphere*: air and noise pollution from transportation vehicles; increased carbon dioxide emission from fossil fuel combustion; temperature variations concomitant with climate change.
- *Water*: introduction of minerals, nutrients, sewage, petrol, and toxins caused by waste disposal; contamination-related loss of water quality.
- *Soil and rocks*: removal of minerals, rocks, fossils, and other specimens; vandalism and graffiti; chemical changes in soil; erosion and compaction.
- *Landscape*: visual impacts of buildings and infrastructure on vistas.
- *Habitats*: decrease in natural habitats due to tree clearing and construction; competition between native and introduced/invasive species; increased fire frequency; changes in germination, establishment, growth, and reproduction patterns; changes in species diversity, composition, and plant morphology; disappearance of fragile species; growth in more resilient species; changes in species composition.
- *Wildlife*: psychological stress, behavioral changes, reductions in productivity due to noise pollution; use of human waste as food sources; avoidance, habituation, or attraction due to increased interaction with humans; changes in species composition and distribution; physiological changes to growth rates and abundance; disruption in feeding patterns; disturbances in mobility patterns caused by roads and barriers; death resulting from traffic collision.

Research on any one of these types of damage arising from tourism is vast. In the way of example we could simply point to the environmental consequences of overcrowding in two of the world's best-known wild places: Mount Everest and the Grand Canyon. In the Grand Canyon the vistas may still be sublime but the soundscapes not so much anymore. In their study of visitors' experiences of noise pollution Mace and colleagues (2004) found that no fewer than 36 different tour operators offered helicopter sightseeing tours above the Canyon. On the busiest days, they reported, up to 100 helicopters could be seen and heard in the airspace above the valley. To be fair, only Mount Logan and Mount Trumbull within Grand Canyon National Park are designated as official wilderness areas, but the broader value of our point remains. Traffic has also become a tremendous problem throughout the park area designated by the Nepalese authorities around Mount Everest. Especially heavy during the peak trekking season in the autumn and spring, traffic (now exceeding 15,000 arrivals per year) has resulted in the overharvesting of alpine shrubs and plants for fuel, corridors of overgrazing, and uncontrolled lodge building (Byers, 2005; Nepal, 2005). Traffic jams have led to further widening of trails and the formation of soil-damaging short cuts (Musa *et al.*, 2004). Increased tourist numbers have also affected waste and sewage disposal and contaminated waters have greatly increased the risk of contracting disease (see Box 6.2 for a worldwide movement that is attempting to counter these trends).

Box 6.2: LEAVE NO TRACE

The concept of "leave no trace" (LNT) refers to outdoor skills focused on leaving minimal to no impact on an environment visited. LNT is an ethic that applies across many outdoor recreation activities and settings. Leave No Trace is also a not-for-profit organization dedicated to the promotion of LNT ethics. The organization's objective is to endorse, encourage, and inspire responsible outdoor recreation through education, research, and partnerships.

With the growing worldwide enthusiasm for outdoor recreation starting in the 1960s and 1970s, designated wilderness sites saw significant increases in usage, which created a heightened awareness of the potential harms of over-visitation (Simon & Alagona, 2013). In 1979, James Bradley of the US Forest Service (USFS) suggested that educational, rather than regulatory, action would be more successful at reducing the negative impacts of backcountry recreation (Marion & Reid, 2001; Simon & Alagona, 2013). Recognizing the tremendous impact that overusage could have on protected areas and parkland, the USFS implemented educational displays and interpretive programs that focused on the concept of LNT (Marion & Reid, 2001). In the 1980s, through collaboration with agencies like the Bureau of Land Management (BLM) and the

Continued

National Park Service (NPS), LNT became more a formalized educational program stressing a new wilderness ethic and the development of seven educational principles focused on educating visitors to minimize the impact on wilderness sites.

In the 1990s, education initiatives continued to prove themselves necessary tools to help prevent damage to heavily visited wilderness sites and parks. The USFS and the National Outdoor Leadership School (NOLS) signed a memorandum of understanding in 1991 and jointly collaborated in developing an educational program (Marion & Reid, 2007). A curriculum was developed with the purpose of disseminating LNT ethics and practices in all aspects of leisure and travel in various environments, ranging from deserts and tundra to waterways, caves, mountains, and alpine forests.

As training programs continued, new and more developed partnerships with BLM, NPS, and US Fish and Wildlife Service (USFWS) emerged. In 1994, Leave No Trace for Outdoor Ethics was incorporated as a non-profit educational organization and is still recognized as the leader in training for educational principles of wilderness preservation ethics. The LNT Center for Outdoor Ethics now develops and distributes a variety of educational materials in cooperation with government agencies, corporate sponsors, and other organizations. The educational material asks recreationalists to: 1. Plan ahead and prepare; 2. Travel and camp on durable surfaces; 3. Dispose of waste properly; 4. Leave what you find; 5. Minimize campfire impacts; 6. Respect wildlife; 7. Be considerate of other visitors (www.leavenotrace.ca/home).

LNT educational programs have had a longstanding impact on how recreationists use and enjoy wilderness sites. Marion and Reid (2007) found that many LNT education programs alter visitor knowledge and effectively encourage visitor behavior that supports the educational principles of the program. However, some critics have also pointed to the ironic contradiction of an educational program that combines conservation ethics with corporate sponsorship and branding. Corporate partners donate to LNT and thus qualify for using the LNT logo for their marketing, communication, and education. Corporate sponsors also have access to LNT principles for promotional material (Simon & Alagona, 2013). Simon and Alagona note that corporate branding and massive consumption of "ethical products" that support LNT philosophy indirectly commodify nature as "leaving no trace in backcountry environments is accompanied by, supported through, and at times dependent upon considerable consumption activities" (2013, p. 333).

Although there is quite a bit of supportive academic literature on the positive effects of LNT education programs, the environmental education program does still show critical reflection. Cachelin, Rose, Dustin, and Shooter (2012) argue that the benefits and principles of LNT are evident in how LNT has been adopted by organizations and groups; however, an issue of concern is that LNT can be counterproductive in encouraging environmentally and socially responsible behaviour. Cachelin *et al.* state: "We attribute this possibility to the prevailing 'humans as apart from nature metaphor' underpinning LNT and recommend it be replaced by a 'humans as a part of nature' metaphor grounded in heightened ecological understanding" (Cachelin *et al.*, 2012, p. 1). The discourse around LNT principles and core values suggests an absence of humans as part of the same ecology that needs protecting. LNT principles protect a place that one visits but does not account for.

Website

www.leavenotrace.ca/home (last accessed December 2, 2015).

From a social and economic perspective—as some of the cases described above have already made apparent—not everyone benefits from ecotourism, and not in the same ways. There are clear symbolic inequalities at play in ecotourist exchanges, for example. As Zanotti and Chernela (2008) have argued, ecotourism-derived education should benefit not only the travelers, but also the locals. In their words, the idea that tourists should be the only ones learning from natural and cultural programs:

> depicts communication between guide/interpreter and tourist as unidirectional, driven by consumer "demand for information" about

> the environment, which the guide/interpreter is to provide. . . . This inappropriately places interpreters in the position of conservation educators, rather than conveyors of information about their own world.
> (2008, p. 497)

For the most part, however, research has concentrated on material inequalities. Amongst the most central and most often-studied concerns are: the potential for "leakage," or the flight of income away from the host country into international companies and investors; the growing economic dependence of host communities on tourism (which essentially works much like a volatile monoculture); the structural inequalities in accessing tourism-generated benefits and revenues; the continued dispossession and forced evictions caused by both ecotourism development and biodiversity conservation; and the loss of tradition and pervasive social change in host communities (for extensive and in-depth reviews see Butler & Boyd, 2000; Frost & Hall, 2009; Honey, 2008).

From a cultural standpoint ecotourism often results in commodifying wild lands and wildlife. For instance, research on wildlife tourism shows that even though activities like birding and wildlife spotting have been widely touted for their potential to finance local economic development, prevent alternative consumptive projects (i.e. logging, mining), and promote conservation initiatives, their beneficial impacts can be difficult to confirm (Lemelin, 2006). Lemelin (2006) in particular has noted that every year more and more wildlife tourist activities become introduced into the worldwide ecotourist market—with suddenly fashionable exotic species becoming subjects of the tourist gaze: from polar bear watching in Manitoba (Lemelin, 2006) to gorilla tourism in Uganda and Rwanda (Litchfield, 2001), and from photographing painted wolves in Kruger National Park (Maddock & Mills, 1994), to viewing Komodo dragons in Indonesia (Walpole, 2001). This has resulted in expanding global spaces of consumption and setting artificial values for some species over others. This process of commodification, Lemelin (2006) argues, has resulted in making the consumptive vs. non-consumptive dichotomy meaningless. "There is little evidence that non-consumptive wildlife tourism activities involve greater empathy, respect or learning benefits," Tremblay (2001, p. 85) also remarks, "it is also potentially liable for large-scale damage to ecosystems, habitats and in the longer run to animals themselves."

Commodification also takes the form of a pervasive loss of authenticity. Brandin (2009), for example, shows how Swedish parks designed for moose safaris have become more and more like zoos and theme parks, with new parks (there are now over 20 in Sweden) designed every year to appeal to a variety of tourist preferences. "Wild," in this sense, becomes a performance emergent from the interaction of people, animals, and specific settings (Brandin, 2009) and ceases to exist as a notion with fixed meaning across different contexts. Communication scholars such as Wiley (2002) provide additional support for this idea. In order to attract tourists, swamp tours taking place in Louisiana manage to construct a sense of wilderness by representing swamplands and their wild inhabitants—from alligators to "hillbillies"—as unpredictable, raw, dangerous, and fundamentally opposed to civilization (all the while ignoring the heavy industrial activities happening nearby). Moreover, commodification directly objectifies wildlife and turns human–animal encounters into a form of shopping, as Lemelin (2006) finds in his ethnographic research on polar bear watching tours in northern Canada. As he shows us, individual animals lose their individual identity when globetrotting nature tourists seek them out as generalized tokens of a species which they simply need to check off their "to see list."

"DRILL, BABY, DRILL!"

In 2008 former Maryland Lieutenant Governor Michael Steele—who was later elected Chairman of the US Republican Party's National Committee—gave a rousing speech in support of the prospect of exploiting the nation's few remaining untapped sources of oil and gas. In the speech's most enthusiastic high note he incited his powerful audience with a crude but powerful slogan which was later adopted by vice president hopeful Sarah Palin. "Drill, baby, drill!"—he screamed. The fact that Sarah Palin so heartily joined the chorus was a scary event for wilderness advocates in the US. Palin, then governor of Alaska, clearly had her sights set on the Arctic National Wildlife Refuge (ANWR): a 77,000 km^2 area of protected land north of the Alaska coast that ranked number one in size in the US's list of officially recognized wilderness areas. This, of course, was not the first time that a politician had lobbied for exploiting gas and oil reserves in ANWR. Since 1977 ANWR has been the subject of a deep controversy between those who wish to protect the grounds of the migrating

Porcupine caribou, and those who insist that developing the 1,500,000 acre-wide zone known as the "1002" area could reduce the nation's dependency on foreign energy.

The ANWR controversy, which arguably will never end, is a perfect example of the "worthless land" hypothesis. Simply stated, the worthless land hypothesis states that parks, reserves, wilderness areas, and similar protected lands are set aside for conservation purposes simply for the reason that they are deemed to be worthless for any other purpose. This idea was first brought forth by Alfred Runte (1979). In his book *National Parks: The American Experience*, he famously observed that in the US:

> There evolved in Congress a firm (if unwritten) policy that only "worthless" lands might be set aside as national parks. From the very beginning Congress bowed to arguments that commercial resources should either be excluded from the parks at the outset, or be opened to exploitation regardless of their location. [Senator] John Conness himself opened the Yosemite debates of 1864 with this assurance: "I will state to the Senate," he began, "that this bill proposes to make a grant of certain premises located in the Sierra Nevada mountains in the State of California, that are for all public purposes worthless . . . The property is of no value to the Government."
>
> (pp. 48–49)

Runte continued:

> The development of the United States in the midst of abundance could not help but strengthen materialism and the nation's commitment to the sanctity of private property . . . . A surplus of rugged, marginal land enabled the country to "afford" scenic protection; national parks, however spectacular from the standpoint of their topography, actually encompassed only those features considered valueless for lumbering, mining, grazing or agriculture. Indeed, throughout the history of the national park idea, the concept of useless scenery has virtually determined which landmarks the nation would protect as well as how it would protect them.
>
> (p. 49)

Runte's "worthless land" hypothesis was set for a wide adoption across the world (Frost, 2004). While we could debate the logic of worthlessness by pointing out that national parks fuel massive flows of tourist money, for now we will simply take away the point that protected areas worldwide are known to face immediate struggles once a group sets their monetary sights on their consumptive potential. Logging, mining, farming, oil extraction, hunting, fishing, construction, and the developments of roads and dams represent the most common threats to wild lands. There is neither the space nor the need to review all the available literature across the social sciences on all these topics. Moreover, these issues tend to receive more interest from news media and political commentators than they do from geographers, it would seem. So, in what follows we will review only a few examples and focus our attention on those that are more theoretically insightful and that offer the most interesting empirical cases.

Let us begin in Ecuador, where a lot of oil sits—untapped—in Yasuni National Park. Yasuni National Park, the traditional home of the Waorani people, is a rich biodiversity reserve located in the Amazon basin, far into the country's northeast. But while rich in terms of biodiversity, Yasuni—and in particular the Ishpingo–Tambococha–Tiputini oil field—is also rich in petroleum. Ecuador's history in dealing with oil reserves has been based on auctioning off concessions to foreign companies, so when in 2007 newly elected president Correa announced that Ecuador would leave oil in the ground of Yasuni National Park the world was caught by surprise. Was this a case of a worthless land remaining worthless in spite of the fact that it had clear financial worth? Not quite. Ingeniously, Correa had a different idea. Since fossil fuel-related emissions are a clear cost to the entire globe, and reducing such emissions is an obviously beneficial service, Correa made a business offer to the world. The oil would be kept in the ground in exchange for $3.6 billion of compensation from the international community (see Davidov, 2012). Plan B, Correa outlined, was to drill. Though the offer was original, if not outright brilliant, things did not work out. As of spring of 2014 Ecuador had only seen $13 million raised against drilling, so oil extraction was officially authorized and set to begin in 2016.

Friends and lovers of all things wild had better luck—if luck had anything to do with it—in British Columbia in the late 1990s. Located on the west coast of Vancouver Island, no more than a couple of hours from our home, Clayoquot Sound is home to massive swaths of old growth forests and large predator species such as black bears, wolves, and cougars. It is also the traditional territory of the Nuu-chah-nulth Nations. As Braun (1997; also see 2002) reports in one of the most influential geographic studies of contested wilderness spaces, however, Nuu-chah-nulth voices were strongly marginalized in a vicious struggle between logging companies and environmentalists. Their marginalization was a postcolonial act, Braun (1997) explains, centered around the definition of who had the legitimate power to speak on behalf of nature. So, while ecologists—who successfully claimed the authority to construct nature and speak for it on their own terms—won the battle and managed to keep the lands largely safe from logging, the debates and discourses surrounding the prospect of logging throughout Clayoquot Sound revealed how social constructions of nature as wild retained their social power only insofar as they excluded human presence and therefore silenced and marginalized indigenous inhabitants.

Just like oil extraction and logging, mining constitutes a powerful threat to wilderness. Recent research has shown mining to be a threat to protected areas in Australia (Aplin, 2004; Hobbs, 2011), Ghana (Hilson & Nymae, 2006), Namibia (Fig, 2008), and Suriname (Haalboom, 2011)—amongst other places. The case of uranium mining inside Namibia's Namib-Naukluft National Park is particularly revealing of the collusion between state and private business interests. As Fig (2008) explains, African governments are under constant pressure from multinational extractive industries and local intermediaries who profit from foreign investment to provide mining concessions, regardless of the existence of biodiversity conservation agreements. One of those agreements protects Namib-Naukluft Park, which is home to one of the world's oldest deserts and the world's highest sand dunes. The park, measured at about 50,000 hectares, is Namibia's largest protected area. The mining site, created in two rounds of gazetting, was opened within the park in 2007, and soon after that the first ten tonnes of uranium went to US firm Converdyne. The share prices of Paladin, the operating mining company, soared by 6,514% within a limited amount of time. Environmental assessment had been contracted out by Paladin to the Johannesburg-based firm Softchem but as Fig (2008) reports and later independent environmental reviews found, numerous procedural and technical shortcomings made the assessment a farce. A nearly complete absence of public debate also characterized the process. In addition critics felt fearful of reprisals and stopped short of speaking out against the project publicly (Fig, 2008).

Roads, at least paved ones, would seem antithetical to the character of wild places. For this reason, many national parks feature road access only to the park headquarters and perhaps to a camping or lodging area, but then keep more remote areas zoned as wilderness free of paved and even unpaved roads. Nevertheless, road access is essential to both tourist activities and consumptive agendas and therefore the prospect of road and even trail construction regularly surfaces as a common threat (for a recent case see Ewah, 2012). For example, as Jacques and Ostergren (2006) report, road access has long been a contentious issue in American national parks and wilderness areas. The Wilderness Act (1964, section 4C) states that "there shall be no temporary road, no use of motor vehicles, motorized equipment or motorboats, no landing of aircraft, no other form of mechanical transport, and no structure or installation within any such area." Yet at times road access is not even the real issue, Jacques and Ostergren (2006) observe, as up to 100,000 snowmobilers whisk through wintery Yellowstone trails every year. Moreover, at times mechanized vehicles are not the problem either as even bicycle and raft access have regularly sparked controversy in the US.

Access roads, as our earlier discussion of the Icelandic case has highlighted, are somewhat of a Trojan horse. Allow something as simple as unpaved road building for safety access, wilderness advocates fear, and much fiercer threats of consumptive activity such as power generation and distribution will follow. In Canada the issue of power generation and distribution in wild and remote areas—areas not necessarily protected but still rich in biodiversity—is at the center of ongoing political debate. The most notable dispute has taken place in the Western Arctic where for over 40 years natural gas industry partners, territorial and federal governments, environmental groups, and First Nations have fought over the fate of the proposed Mackenzie Valley Pipeline, a project which would channel flows of natural gas from the Beaufort Sea all the way to Alberta. Though currently under construction after a very long process of environmental assessment, the project was initially scrapped following an inquiry led by Justice Thomas Berger. Berger's 1977 negative decision, driven by the need to respect First Nations' stakes and rights, was considered to be monumentally important for the time and has been deeply influential for all cases that have followed. Nonetheless, pipelines continue to threaten wild areas. The latest threat is the Enbridge Northern Gateway Pipeline: a 1,177 km pipeline that would transport natural gas condensate eastbound and diluted bitumen westbound between Bruderheim, Alberta, and Kitimat, on the British Columbia coast—where tanker ships could virtually stretch out the greasy link throughout the rest of coast and across the Pacific. But in this case in June 2014—in spite of opposition by First Nations, as well as many citizen groups, local municipalities, and ENGOs (environmental non-governmental organizations)—the conservative federal government approved the project.

Traditional consumptive activities like fishing, hunting, and farming are also deeply exploitative of wild places and, obviously, wildlife. On the surface, one of the most heinous forms of exploitation is trophy hunting: a highly controversial activity typically consisting of rich men paying hefty fees to be licensed to kill large animals from the comfort of their safety positions and through the use of their overpowering weapons.[1] While our personal attitudes about trophy hunting are quite obvious, the case is greatly complicated if we take into account the economics of it all. Managed—that is, controlled—trophy hunting can be sustainable if consumer costs are kept consistently high (the trophy hunter market is notably inelastic, as opposed to the more sharply fluctuating middle-class consumer ecotourism market) and the hunting quotas low. Through proper management the revenues from trophy hunting can then be reinvested in conservation efforts, the meat can be salvaged, and predator population can be kept at ideal numbers. While animal right activists argue that trophy hunting is unethical, well-managed trophy hunting can also reduce poaching and fuel social and economic development (see Darkey & Alexander, 2014). As Darkey and Alexander point out, however, the quota-setting process and the whole management of trophy hunting is regularly subject to corruption and easily amenable to manipulation and regulation infraction (Darkey & Alexander, 2014).

Arguments over what constitutes "sustainable" fishing, hunting, and farming happen every day, and everywhere the vocabulary of sustainability has taken root, but in the meantime exploitation—sustainable or not—continues, as several cases from the current research literature show. For example in Antigua, as Day (2007) reports, agriculture, quarrying, and tourism are putting tremendous pressure on the fragile and rare karst environment. Small national park sites at Great Bird Island, Devil's Bridge, Jabberwock Beach, and Bat's Cave have gone some way toward offering protection, but of the 110 km^2 of karst only 10 km^2 have been granted conservation status. Farming has also been reported to be problematic at Cuc Phuong National Park, in Vietnam (Rugendyke & Thi Son, 2005). When the park was created its indigenous inhabitants were resettled to a buffer zone outside the park boundaries, but the question of their new livelihood was not properly addressed. Resettled people therefore had no choice but to hunt birds, and raise chickens, pigs, and cattle inside the park—a problem known to many protected areas worldwide.

Illegally farming inside a protected area, however, is not always as pernicious a problem as that of legally farming outside of it and yet still negatively affecting ecologies inside it. Take salmon farming, for example. For many years now the Norwegian salmon industry has conducted aquaculture right alongside areas still inhabited by wild salmon. Even though farmed salmon are raised in large cages the water they share with wild salmon is virtually the same. This has generated a great deal of controversy since it has been found that sea lice (*Lepeophtheirus salmonis*) become highly concentrated in salmon farms and then tend to propagate outwards, negatively affecting wild salmon populations. In cases like this, the boundaries between industrial/farming activities (salmon farming) and wilderness become fully permeable.

Moreover, as Lien and Law (2011) argue, other important boundaries become blurred. In Norway the word "*villaks*," wild salmon, has only been recently added to the Norwegian vocabulary. That is because prior to the establishment of salmon farms salmon was simply "*laks*," that is: salmon. Lien and Law (2011, p. 74) explain:

> Thus, the new term evolved with the emergence of the young salmon industry, and served to distinguish farmed salmon from that which was the preferred catch of anglers and net fishermen in rivers and fjords. Hence, the term "villaks" is relational, defined in contradistinction with domestication, and draws on the idea of wilderness. It re-enacts a particular version of nature that has its specific historical origin in European thought, and that idealised nature as a non-human realm (see also Cronon 1996a). Specifically, it performs a distinction between non-human nature (wilderness) and society. The prefix "vill" ("villaks") establishes a kind of guarantee that this salmon is the "real thing."

Nevertheless the new notion of "villaks" obscures important historical and social realities. It hides, for example, the fact that Norwegian fishermen have hatched, cultivated, and released salmon fry in order to manage their population since the nineteenth century (Lien & Law, 2011). The notion of "wild" in this case obscures the fact that (1) wildness is a relational entity borne out of the establishment of its very opposite and (2) partly domesticated species have long been falsely believed to be wild, free of human intervention. This realization, we feel, is a very good place to end this chapter and launch our conclusion. We exit by quoting again from Lien and Law (2011, p. 83) who view "villaks" as revealing much broader cultural and social dynamics:

> So what do we learn about nature? Most straightforwardly, we learn that it is being done, done again, and done again in the fish-related practices, personal, economic, regulatory, and scientific, of a country such as Norway. We learn that, like salmon, it shifts its shape and form from practice to practice. It is done multiply. Does this mean, then, that the nature–culture divide is no longer foundational? The answer depends on what we mean by foundational. If foundations are invariant and immovable structures, then the answer is: these do not exist. If, on the other hand, foundational dichotomies are forms that reappear, in different but related ways in endless practices, then the answer is yes: these do exist. The nature–culture divide is messy, it is heterogeneous, it is complex, and it is not coherent. But it is endlessly consequential for everyone involved in fishy practices and other forms of relations that involve non-human beings. A performative approach to nature practices is one way in which we can begin to understand these consequences.

SUMMARY OF KEY POINTS

- Regardless of what we do in wilderness areas, by virtue of our mere presence we all engage in exploitation.
- Exploitation can be consumptive (e.g. fishing, hunting, logging, farming, mining, construction) or non-consumptive (e.g. ecotourism). However, non-consumptive activities are often very consumptive indeed.
- Ecotourism, despite its benign intent, suffers from practical challenges that make its ideals very difficult to respect.
- Key ecotourism challenges include social, cultural, and environmental ones.

- Consumptive activities present key environmental threats to wilderness areas, yet their appeal consists in the potential to generate economic gain. The "worthless land" theory shows how areas remain protected only insofar as no other uses for them are foreseen.
- Consumptive activities may be fair and equitable when multiple stakeholders reach agreement over planned activities, and when environmental sustainability is given careful consideration.

DISCUSSION QUESTIONS

1. Can ecotourism ever live up to its ideals?
2. In what ways is ecotourism consumptive?
3. What advantages do countries derive by marketing themselves as ecotourist destinations?
4. In what ways do ecotourist bubbles work?
5. In what ways can local communities derive greater benefits from ecotourism?
6. Are ecotourists different travelers than mass tourists? How so?
7. In what ways are wildlife tourism, ecotourism, and nature-based tourism similar and different?
8. Can consumptive activities like mining ever be sustainable?
9. Is sustainable hunting nothing but a case of greenwashing?
10. Are fish farms necessary evils to feed a growing world's appetite?

KEY READINGS

Braun, B. (2002). *The intemperate rainforest: Nature, culture, and power on Canada's west coast*. Vancouver: UBC Press.

Frost, W. & C. M. Hall (eds) (2009). *Tourism and national parks: International perspectives on development, histories, and change*. New York: Routledge.

Honey, M. (2008). *Ecotourism and sustainable development: Who owns paradise?* Washington, DC: Island Press.

WEBSITES

The International Ecotourism Society: www.ecotourism.org

Sustainable Tourism Online: www.sustainabletourismonline.com

Hunting for sustainability: http://fp7hunt.net/

African Wildlife Foundation: www.awf.org
United Nations Environment Programme: www.unep.org
International Rivers: www.internationalrivers.org

All of these websites were last accessed on December 2, 2015.

NOTE

1 For a recent discussion of this practice in the news see "The death of Cecil the Lion" (2015) in the *New York Times*.

7

REASSEMBLING WILDERNESS

Assembling the river, China (Photo: April Vannini)

Since this book began with a walk in the woods—as you might recall from the opening words of Chapter One—it is fitting we should end it by returning it there. It is summertime now and our island's forest looks, smells, and sounds different than it did a few seasons ago. Though it is only early June the grasses are already yellow and withered, thanks to an incredibly long spring drought. The soundscape is remarkably different too; summer visitors have arrived and it is no longer a reasonable expectation to lose sight and sound of humankind through a short jaunt in the forest's trails. The smell of dampness is gone too. Now the air smells dry and the scent of pollen has replaced that of wild mushrooms. In spite of all the natural change and the momentarily higher density of people, the character of the anonymous "bush"—which had prompted us to think of the concept of wilderness as being more than simply a legally protected area worthy of official wilderness categorization—still feels very much the same to us, as wild as ever.

Nonetheless, six chapters and hundreds of cited resources later, we feel we can no longer afford to get away with a flippant "wilderness is as wilderness does" kind of answer to our research question. Surely, as ethnographers, our emic approach—consisting of finding wilderness where others have claimed it exists—is convenient, but by now we should at the very least try and be more descriptive and arrive at some sort of conceptual inventory that captures where in the world, and why, people might think they run into the wild. Such is the purpose of this chapter.

What we will do in the pages to follow is engage in a rather simple argument: the argument that wilderness is an assemblage. Though we will endeavor to duly deepen this idea with an arsenal of conceptual weapons, theory, and some research, our claim will always be easy enough to boil down to the notion that wilderness is something "thrown together" by a multitude of actors. We will refer to these actors' activities as a kind of "assembling" and to the resulting outcome as an assemblage or, as we will later specify in order to be more accurate, a meshwork. Unlike other critics, we do not believe we should purge the English dictionary of the word "wilderness" and replace it with something more fashionable, like "wildness" or "biodiversity" (e.g. Callicott, 2008). Sure the idea of wilderness can evoke some negative feelings in some circles, but so do words like holocaust, war, or genocide, and by not speaking them we obviously cannot make their referents go away. Wilderness, we believe, simply needs to be better understood. If by continuing to refer to it we must recall its tumultuous heritage, then we are all the better for acting mindfully, responsibly, and reflexively.

By treating wilderness as an assemblage we do not mean to re-hash the idea that wilderness is a social construction. We need a better metaphor, one that de-emphasizes the somewhat instrumental and arbitrary notion of "constructing"—which can perilously slide into an idealist notion of making anything out of just about everything. The concepts of assembling and assemblage can usefully align our treatment of wilderness with recent developments in geography, developments that have called for the growth of a "more-than-human" treatment of the world which takes into serious account the materials that comprise the lifeworld, our entanglement with such materials, and our role as humans in the world's ongoing transformation of itself. Much of what we argue in this chapter has been duly elaborated elsewhere, chiefly in assemblage geographies (Robbins & Marks, 2009), in relational orientations to the environment (Braun, 2006), and in theoretical developments that emphasize inhabitation, rather than construction, of the world (Ingold, 2010, 2015), so we merely apply existing ideas. We begin with a brief explanation of what an assemblage is and then delve in greater depth into how wilderness sites are assembled. Then, we shift our discussion to the notion of the meshwork. After a brief review of what meshworks are we discuss how wilderness is knotted together as a meshwork.

WHAT IS AN ASSEMBLAGE?

Assemblages can be formally defined as "wholes characterized by relations of exteriority" (DeLanda, 2006, p. 10): a succinct but admittedly cryptic definition that will require a lot of unpacking. Such unpacking is the purpose of this section. We should start by mentioning that the notion of assemblage, as understood with contemporary geography and other social sciences, owes a great deal of conceptual debt to philosophers Deleuze and Guattari (1987). Writing in their native French, Deleuze and Guattari, however, never quite wrote about assemblages but rather about *agencement*: a word without a clear translation in English (Markus & Saka, 2006; Phillips, 2006). After the worldwide diffusion of their work, *agencement* became "assemblage" in English, thus losing some of its original meaning and becoming something else (and, ironically, in doing so confirming the validity of their ideas on emergence, as you will understand shortly!). Regardless of whether one prefers to remain faithful to the original or its derivation (it is the latter strategy we follow) the "relations of exteriority" that we mentioned refer to the fact that "component parts of a whole cannot be reduced to their function within that whole, and indeed they can be parts of multiple wholes at any given moment" (Dittmer, 2014, p. 387). The functions and capacities of all these various parts, however, are deeply shaped by their finite properties and their infinite relations with the whole. Assemblages, therefore, can be understood as complex interaction systems, complex ecologies. But let us proceed slowly and introduce a couple of simplifying examples.

You might have heard of, or seen first-hand, assemblage art. Assemblage art is quite different from representational art forms like, say, landscape painting. A landscape painting shows us a panorama, a place, a view in quite a realistic fashion, as its main aesthetic intent is to mimic, to represent, and to depict (we are aware we are simplifying things quite a bit, but do bear with us). On the other hand assemblage art aims to build something new, to throw something together, and to knot disparate pieces which together create something novel in virtue of the unique relations that bind them. Artistic assemblages, arguably much more than representational art forms, thus challenge our curiosity as to how something works by forcing us to disentangle how, precisely, they function, how they have come to be what they are, and what they do and could potentially do if we reassembled them in different ways. Wilderness is a bit like assemblage art: the naming and entanglement of all its various parts is a creative and imaginative human act, but it is one that could not work if it were not for the capacities of their different components and their potential for becoming knotted together with each other. So we want you to think of wilderness sites as less like representational art and more like something like assemblage art (again we do apologize for the simplicity of our contrast) and therefore less to be judged against a normative ideal (e.g. the idea of the "pristine wild"), and more to be beheld for the capacities of their unique constituent parts, how they work together, and how they could work otherwise.

If you are not familiar with assemblage art, the next two examples might work better for you. The first, as introduced earlier in this book, is that of an automobile. An automobile is a set of relations of different parts with one another. By tracing connections among the parts we view tires, windows, pistons, a battery, and a fuel tank not so much as units in themselves but for what they make up together. "Assemblage," McFarlane and Anderson (2011, p. 162) observe, "functions as a name for unity across difference, i.e. for describing alignments or wholes between different actors without losing sight of the specific agencies that form assemblages." So obviously the resulting assemblage, the car, could not be what it is without an organizing act: an agentic practice which we might call an act of assembling. That act of assembling could always have unfolded in different ways since socio-spatial formations like assemblages are always open to the possibility of being otherwise (McFarlane & Anderson, 2011); however, assembling is not an arbitrary act of social construction, or of attaching meaning to an inert object. Material and discursive constraints, as well as the peculiarities of their different historical trajectories, limit which assemblages take hold of the imagination and which ones do not.

Another handy example of an assemblage might be that of a city bus route. Public transit planners create bus lines by connecting together different urban neighborhoods and their various streets. Bus lines therefore comprise specific sites (like bus stops) but also a myriad of other components (bus stop signs, benches, schedules, buses, drivers, roads, regulations, fare systems, etc.) that bind the bus line assemblage together with a multitude of other urban assemblages (airports, train stations, shopping centers, etc.). What is interesting about this example—more so than the automobile example—is how bus routes are territorialized at the moment of their conception, but also how they are constantly subject to reterritorialization. Think for example of how special events, road construction, traffic accidents, residential and commercial development over time, commuter appeals for schedule or route changes, and so forth "cause" change to the way a bus moves along a line over the years. Acts of assembling and reassembling belie the constant tug of war between stability and change, organization and rupture, balance and tension (McFarlane & Anderson, 2011). However, rather than viewing these actors' practices as *causes* for change, and the resulting changes to the system as *effects* that radically alter the original identity of the bus line, we should view the activities of all the components of this assemblage as ongoing manifestations of their open-ended capacity to act, their ability to grow. Therefore, we should think of change, growth, and transformation not as corruptions of an initial entity's form, but rather as evidence of its aliveness. In other words: we cannot think of assemblages apart from their openness to change and the endless actualization of their potential to transform their relations. Thinking of wilderness this way should help us in keeping in mind that wilderness is not a pristine place frozen at degree zero of development; that is, a place that cannot accommodate any change.

As these examples have shown, assemblages are particularly apt at describing how places are subject to processes of formation and transformation. The idea of assemblage introduces into our thinking about place a "radical sense of openness and possibility" (McFarlane & Anderson, 2011, p. 162) which emphasizes the recombinant diversity of its parts without discounting their specificity and unique qualities. How precisely change occurs within assemblages is, of course, the crux of the matter. No definitive generic answers exist, so the name of the game is to describe and interpret such change. Change could occur as a result of intensification happening in certain parts and relations among parts. Or it could surface as existing components become entangled in new arrangements. Or perhaps change could occur as forces outside an assemblage disrupt relations within it (McFarlane & Anderson, 2011). All of these are essentially questions awaiting study on a case-by-case basis.

Social theorist Manuel DeLanda (2006) explains that a particular assemblage is delineated from neighboring assemblages through dynamics of territorialization and deterritorialization. Territorialization can be loosely thought of as a movement toward internal organization and coherence, while on the other hand deterritorialization describes forces which rupture and unsettle the whole. As well, dynamics such as coding and decoding work toward order and chaos, either by respectively consolidating and rigidifying an assemblage or alternatively by introducing flexibility into the system (DeLanda, 2006, p. 19). All of these dynamics, Deleuze and Guattari (1987) observe, are therefore characterized by "lines of flight": essentially open-ended trajectories of change that make a wide range of futures possible for assemblages. Of course not all futures are actualized, but their potential to become real makes our work as students of assemblage highly receptive to the need for anticipating change. A student of assemblages, to put it in simpler words, therefore need not worry as much as others do about the boundaries between authenticity and inauthenticity, originals and their derivation, and the incipient and its evolution.

Another advantage inherent in using the idea of assemblage is its usefulness in describing more-than-human entanglements. Assemblages combine human and non-human actors, organic and inorganic materials, discourses and technics. Assemblages knot together "politics, machines, organisms, law, standards and grades, taste and aesthetics, even the production of sovereign territory and the politics of scale" (Braun, 2006, p. 647). By remaining open to the possibility of multiple rearrangements and by putting a premium on the plurality of capacities of its components, assemblages move beyond the "notion of Nature as singular and universal" (Braun, 2006, p. 644). Now that we have outlined what assemblages are and how practices of assembling can impact assemblages, let us abandon the abstract nature of our discussion and move on to discuss how wilderness sites are assembled.

WILDERNESS ASSEMBLAGES

Wilderness is an idea, and in fact an ideal. It is also a place. Wilderness, or at least wildness, is also a quality, a character, an atmosphere. Yet wilderness is also a very concrete legal category. Wilderness, to get to the point, is a complex entity with multiple facets which take many different shapes in different geographical, social, cultural, political, and historical contexts. Its elements include non-humans (trees, rocks, animals, documents, fences, maps), socially situated human subjects (visitors, legislators, stakeholders, law enforcers), and objectives (profit, sustenance, scientific data collection, leisure, conservation), as well as a diverse array of bodies of knowledge and discursive practices (laws, regulations, policies, scientific information). Each ensemble of these components makes up a particular wilderness site, with each site somewhat similar but also distinct from others. How are different wilderness sites assembled then? In order to disentangle assemblages, Robbins and Marks (2009) suggest, we should focus our attention on three things: (1) the relationships between humans and non-humans; (2) the nature, directions, and rhythms of change; (3) the spatial dimensions of an assemblage. But while this is a good start, we need some more precise tools, tools that can allow us to outline the work that goes into drawing heterogeneous components together and cementing and sustaining coherent connections. In what follows we adopt Li's (2007) list of six generic assembling practices: forging alignments, rendering technical, authorizing knowledge, managing failures and contradictions, practicing anti-politics, and reassembling.

First in Li's (2007) list of practices to examine is *forging alignments*, or in other words the work that goes into combining together the various objectives of the different parties with a stake in a particular assemblage.

This includes parties who aspire to actively define or regulate an object and also the groups and individuals who hope to benefit from such definitions and regulations. For example, officially defining a site as a wilderness area or otherwise as a protected area with a wild character may require forging alignments between local residents, land owners, visitors, businesses, the local workforce, local lawmakers and governments, law and regulation enforcers, state institutions, conservation activists, and scientists. At times only some of these parties may be involved, at times all may be. In certain circumstances some of these groups or individuals may be notably more powerful than others, whereas in other situations the balance of power may be more evenly distributed. Objectives may also differ from case to case, with some cases being characterized by more widely shared visions and other cases being marked by disparate goals. Regardless of the particulars, a focus on forging alignments teaches us how individual wilderness sites are intersubjective accomplishments subject to negotiation and compromise.

The second practice (in no particular order) requiring our attention is *rendering technical* (Li, 2007) or, arguably in better words, the practice of operationalization. By "rendering technical" Li (2007) refers to how assemblages are drawn out of a disorderly reality and thus rendered precise, instrumental, and operational. Operationalization comprises of processes that specify the exact socio-nature of a wilderness and make it subject to problem-solving intervention, management, and monitoring. This is especially the case for wilderness areas that are the explicit subject of strict regulations, such as IUCN Category 1b areas that are legislated and administered by state authorities. For example, writing about Russia's Karelian Old Growth Forests, Kortelainen (2010) outlines how essential in the formation of an old growth forest is the calculation of a wooded area's boundaries, a census of its ecological resources, and a definition of its "nature" in very precise detail. This operationalization may require on-the-ground ecological research, satellite technology, cartography, etc. Operationalization, in sum, is a process delegated to technical experts who by the nature of their work turn to a space as a knowledge management problem. Their activities in turn influence later assessments by additional experts called upon to treat a wilderness site as a unit of analysis and the subject of technical administrative tasks.

Authorizing knowledge is another practice by which wilderness sites are assembled. As Braun's (2002) research has shown in great detail in the case of Clayoquot Sound, *who* is authorized to speak on behalf of nature is a process subject to various processes of legitimization, institutionalization, and contestation. By gaining the authorized knowledge to shape the destiny of a wilderness site a particular human agent can exert the force of a specific body of preferred knowledge. Such positionality gives authorized knowledge holders the power to fend off critiques and pursue their objectives in an authoritative way. For example, by tracing their legitimacy to traditional custom and to historical patterns of inhabitation aboriginal groups worldwide have been able to claim lands as their own, utilize traditional ecological knowledge to use and protect places, and become active managers and regulators of wilderness areas (often rejecting, in the process, the moniker of "wilderness"). When we view wilderness sites or wild places as the relational outcomes of knowledge systems, however, we must not forget to take into account how those with the authorized knowledge to assemble wilderness are in turn shaped by the wilderness assemblages they have territorialized. Thus, for instance, the very notion of traditional ecological knowledge and its application has resulted in the production of the myth of the environmental noble savage or the "Ecological Indian" (Krech, 2000).

Li's (2007) fourth assembling practice is *managing failures and contradictions*. This practice allows wilderness stakeholders to maintain stability in the face of challenge by minimizing threats through compromise, reconceptualization, redefinition, amended policies, and changes in management and in regulation, as well as shifts in leaderships and alliances. We should view the handling of contradictions as more than a managerial problem. Take, for example, the case of a private piece of land commonly regarded as wild which is suddenly developed to make space for an ecotourist resort. Does the development render the previously "pristine" environment something other than wilderness? Does the resort's presence, together with its ecological sustainability mandate, turn a previously nondescript "bush" into a more widely known wilderness site? Answers to these questions, the resulting debates these issues originate, and the ideological struggles which end up informing the meaning of "wilderness" and "wild" are unstructured or only loosely structured ways in which a particular wilderness assemblage manages contradictions and reaches coherence—albeit always in a tentative manner constantly subject to further stress and disruption. As Kortelainen (2010) discusses, the concepts that inform our understanding of nature and wilderness are fluid and continuously subject to processes of adaption and conversion to local circumstances. Think, for example, of the notion of a minimum size of a wilderness area and how failures to protect large swaths of land have given way to alternative, yet coherent and widely accepted, smaller conservation programs.

Wilderness areas are contested territories, as we have seen throughout this book. As Hobbs (2011) finds, conservation-minded agents and developers are often at odds with each other over the "meanings" of specific wild lands. These environmental disputes are battled through multiple personal experiences and perspectives, ideological positions, competing knowledges, and stories that are "performed, contested, re-performed and reinforced on the public stage" (Hobbs, 2011, p. 127). These disputes are what Li (2007) refers to as *anti-politics*. Anti-politics works by way of attempting to close down debate about how to govern a territory—typically by making reference to expertise and authority. Anti-politics, despite the name, is fully political: a process through which different parties invoke and enact the validity of their preferred vision of wilderness. The quintessential element in wilderness debates, Hobbs (2011) argues, is the environmentalist claim that wilderness transcends human interaction and that therefore it has always stood and should continue to stand outside of politics. In this way, however, the "pristine" character of wilderness becomes the subject of endless contention, with developers often attempting to show evidence of longstanding human presence and resource utilization. Anti-political processes aim to shut down these claims and their respective proponents.

A sixth practice we should examine is that of *reassembling*, which consists of reorganizing and reordering assemblages by "grafting on new elements and reworking old ones; deploying existing discourses to new ends; [and] transposing the meanings of key terms" (Li, 2007, p. 265). In her analysis of community forest management, for example, Li (2007) finds that since the 1990s shifts in forest management have been informed more and more by neo-liberal discourses. These discourses and their related practices have resulted in downsizing and decentralizing the administration of forests and emphasizing the value of forest-based entrepreneurialism and market efficiency. More broadly, as we have seen in Chapter Six, wilderness is increasingly subject to similar neo-liberal processes as well. Ecotourism, community resource management, zoning, and similar schemes and agendas have in many cases turned wilderness areas away from the fortress model of conservation and fully reassembled wilderness as a utilitarian space.

Even in those cases where wild lands have been protected from development and restricted to low-impact recreational use we are now seeing more and more corporate interests at play, for instance in the outfitting of sponsored expeditions and in their documentation and broadcasting for distant entertainment purposes. Through these processes wilderness is more and more commonly assembled as a playground for those with disposable income and available time and skills.

Now, the six practices we have discussed above are a good start, but by no means is this list all-inclusive. Many additional practices through which wilderness is assembled could be listed, and it is only through in-depth empirical research that more questions about how wilderness is done can be generated and more sensitizing concepts coined. We should also be very reflexive about the nature of these practices before closing down this section on assembling and assemblages. It is obvious that the six practices outlined above are very practical and useful in dissecting how wilderness is done. In fact, we should remember how in Chapter One we set out to do precisely that: to practically disentangle how we perform wilderness into being—rather than to debate what wilderness is, or where its essence lies. Nevertheless, while they are a good start these six assembling practices do not go far enough. Though easy to study, their focus is quite limited to the discourses and actions of human actors. Anyone studying wilderness assemblages by teasing out these generic social processes should quickly come to the realization that the ties which these practices bind are rather social, cultural, and political in a traditional sense. A material, biological, and place-based understanding of wilderness assemblages is required if we wish to fully understand more-than-human wilderness assemblages from a more ecological viewpoint. So, what to do next?

A possible way forward is to better stress the relational way in which assemblages come to be by further investigating how people, things, and the broader material world of substances, movements, and forces are enmeshed with one another. There are different methodologies and different theoretical orientations and analytical perspectives we could apply to this task but we believe that a particularly useful approach is the application of an alternative metaphor. Metaphors allow us to juxtapose alternative mental images, to apply these images to a particular case, and thus to reveal novel patterns which would have remained invisible had it not been for the use of such metaphors. Metaphors, in simpler words, are great tools to imagine, think, learn, and do things with. Metaphors are particular useful to our intents and purposes if they allow us to transcend the limitations of traditional concepts of the "social." Traditionally, by "social" the social sciences have referred to the binds that tie different people together. But if we wish to embrace a more-than-human perspective of relationality then "social" must be intended to encompass relationships based on entanglements among different classes and types of actors, both human and non-human. Furthermore, these metaphors allow us to leave behind dated distinctions between the social and the natural, thus revealing how wilderness is done through the entanglement of socio-natural forces and actors.

In their discussion of assemblage geographies Robbins and Marks (2009) outline four alternative metaphors for our understanding of assemblages: symmetrical, intimate, metabolic, and genealogical. Briefly (see Robbins & Marks, 2009, for more detail), symmetrical assemblages draw from the theoretical ideas of actor-network theorists such as Bruno Latour. Intimate assemblages are chiefly informed by posthumanist thinkers such as Donna Haraway. Metabolic assemblages are mainly inspired by Marxist and post-Marxist ideas. And genealogical assemblages owe their inspiration to the thought of Michel Foucault. All of these metaphors are valuable, but we prefer to deepen our understanding of assemblages by turning to the post-phenomenological and non-representational perspective embraced by thinkers such as Tim Ingold, who employs an altogether different metaphor throughout his later work (e.g. 2010, 2015): the meshwork. In the next section we introduce Ingold's idea of the meshwork, and subsequently we endeavor to apply it to the study of wilderness.

THE MESHWORK

What is a meshwork? Dictionaries define a meshwork as an open fabric made of either string or rope that is woven or tangled together at regular intervals. For our metaphorical intents and purposes, however, any material will do and the intervals and the patterns need not be so regular or symmetrical; a meshwork is, essentially, a tangled web. A meshwork, let us be clear from the onset, is not a network. As Ingold (2010) explains, the meshwork is made and constantly remade by tangled lines of growth and movement. The network refers to something related but also quite different. The key difference, Ingold (2010, 2015) argues, lies in the openness of the meshwork. Network-based metaphors have a tendency to divide the world into points of contact among connected nodes whereby something is either inside or outside the network. The meshwork metaphor, on the other hand, puts emphasis on a world in constant formation, an open world that is constantly unfolding and moving. Let us discuss this in greater depth by beginning with Ingold's important critique of the logic of inversion.

According to Ingold, Western worldviews are characterized by the logic of inversion, a logic that is so deeply entrenched in our ways of thinking and acting that it is actually quite difficult to become sensitized to it. "Through this logic," Ingold (2010, p. 68) explains:

> the field of involvement in the world, of a thing or person, is converted into an interior schema of which its manifest appearance and behaviour are but outward expressions. Thus the organism, moving and growing along lines that bind it into the web of life, is reconfigured as the outward expression of an inner design. Likewise the person, acting and perceiving within a nexus of intertwined relationships, is presumed to behave according to the directions of cultural models or cognitive schemata installed inside his or her head. By way of inversion, beings originally open to the world are closed in upon themselves, sealed by an outer boundary or shell that protects their inner constitution from the traffic of interactions with their surroundings.

In exemplifying how the logic of inversion works Ingold (2007) points our attention to how we typically draw a fish. When asked to draw a fish most of us would typically draw a roughly oval shape, add a triangle-like tail, then a fin above and below the body, and finally a couple of beady eyes and mouth. We think nothing of the act of drawing a fish this way; that's what a fish looks like in our minds. By doing so, however, we reveal the logic of inversion at work. A thing, in this case a fish, is converted into a schema that reduces its "being" to its outward appearance and physiological boundaries with the outside. Inside the drawn shape we thus find fishness. Outside of it we may find water, air, the kitchen table, or something else; something other than fish and its fishness. In this way a being that is originally fully immersed in the world becomes closed in upon itself, sealed by the edges drawn by our pencil and our cognition, and fully dissected from the currents of its entanglement with the world.

"Fine!" you might say, "I am not an artist, and that's as well as I can draw." Fair enough, but why represent (and think, and imagine, and know) a being in virtue of its boundaries, rather than in virtue of its entanglements, its movements, and its becoming? Why not draw, for example, a zigzag to denote the fish's swimming and a few wavy lines on top and below to denote the flow of the currents it swims in (Ingold, 2011)? That kind of thinking and imagination is precisely what the metaphor of the meshwork invites us to engage in.

The meshwork metaphor, it should be obvious by now, invites us to put the logic of inversion into reverse (Ingold, 2010, p. 68). Meshworks force us to confront life as becoming, as movement, as something entangled in multiple currents of formation. Meshworks invite us to treat life as lived in the open, in a world that is not pre-constituted or occupied by things existing independent of other things. Ingold (2010) argues that to fully appreciate the power of the meshwork metaphor (and arguably in his thought the meshwork is not so much a metaphor, but rather an essence of sorts, or a core image) we need to understand how life is experienced and practiced as a form of movement. Movement is the quintessential manifestation of existence. Movement generates relations and thus weaves the knots that bind things and people together. And movement, according to Ingold (2010, 2015), cannot be understood outside of the traces—that is the "trails"—it takes place in and creates. "Neither beginning here and ending there, nor vice versa," he explains (Ingold, 2010, p. 71):

> the trail winds through or amidst like the root of a plant or a stream between its banks. Each such trail is but one strand in a tissue of trails that together comprise the texture of the lifeworld. This texture is what I mean when I speak of organisms being constituted within a relational field. It is a field not of interconnected points but of interwoven lines; not a network but a meshwork.

Now, are metaphors such as the meshwork unique types of assemblages? Or are assemblages and meshwork different and therefore incompatible ideas? In his later work Ingold (2015) quite explicitly criticizes assemblage theory and the idea of the network (which is quite popular within assemblage geographies) for depending on a principle of separation. "The network metaphor logically entails that the elements connected are distinguished from the lines of their connection," Ingold (2010, p. 70) explains, and that "the establishment of relations between these elements—whether they be organisms, persons, or things of any other kind—necessarily requires that each is turned in upon itself prior to its integration into the network." Such principle of separation is typical of the logic of inversion and therefore stands in contradiction of the principle of openness characteristic of meshworks. For our part we leave it up to our readers to determine which elements of the different metaphors are more or less useful to them in order to understand how wilderness is done. There is a lot of value in assemblage geographies, and while some practitioners of actor-network theory may indeed employ concepts or images that do not cohere well with the meshwork metaphor, what matters to us is that as long as we keep in mind that "things *are* their relations" we are well on our way to understanding what matters (Ingold 2010, p. 70). Let us now move on to applying the metaphor of the meshwork to our study of wilderness.

WILDERNESS AS MESHWORK

Earlier in this chapter we have outlined how different actors are entangled in wilderness assemblages. Even though meshworks and assemblages are not the same thing, we believe they are related and many of their common principles work in similar ways. Thinking of wilderness as a meshwork, however, pushes our imagination farther. Proceeding with Ingold's ideas we could start envisioning wilderness as knot of intersecting lines, a mental image that could very well be used to describe the idea of wilderness as an ideal, or a particular wilderness area, a single tree, or mountain, or forest, or forest dweller, or temporary visitor. "Organisms and persons," Ingold (2010, p. 70) explains, are "knots in a tissue of knots, whose constituent strands, as they become tied up with other strands, in other knots, comprise the meshwork."

"Fine, I get it, wilderness is like a meshwork" you might counteract, "but what good is a metaphor for showing me what wilderness is if it also applies to feral goats, wild salmon, fisher folk, kayaks, and everything else in between?" Well, that is a fair question, but it is founded on an erroneous principle: for us what wilderness *is* actually matters less than understanding how wilderness is *done*. And for that, the meshwork is very useful. Let us examine how.

Let us return to square one once again: a simple walk in the woods in search of wild mushrooms. Imagine that after a half hour or so of walking around the forest you finally lay your eyes on a prized bolete (*boletus edulis*): one of the meatiest, most flavorful wild mushrooms there is (you might know them as "ceps" or "porcini" perhaps). Excited at its sight (you haven't found one in a while because they haven't been in season) you kneel down on the ground to take a good whiff at it. And as often happens, as you change your angle of sight you spot another one, and then a smaller one too. "This is awesome!" you shout, as you frantically reach for your Swiss Army knife. You decide to keep the smallest one in the ground but cut the other two, making sure to cut them toward the bottom of the stem as to keep the "roots" into the ground. Speaking of roots, if you were to dig them up (which you shouldn't) you'd get a good peek at one of the clearest examples of a meshwork: the fungal mycelium (Google Image that word to get a sense of what it looks like). With their multiple tangled lines, mycelia are the "prototypical exemplar of a living organism" because they show so vividly how an organism is enmeshed in its environment and how its life, its becoming, has unfolded as a series of traces of its movements (Ingold, 2008, p. 1807).

If you manage to pull your attention away from mushroom spotting you might even notice a few other entanglements. Above your head and right around your body tree branches, shrubs, and bushes have reached out to one another through their twigs and leaves. The tree canopy itself, you will realize when you lift your head to gaze right into the sky, is composed of large branches and trunks that lean into and then away from each other, then into each other and then away again, as the wind pushes them and pulls them apart. Below your feet, more tangles. Ingold reflects:

> What we have been accustomed to calling "the environment" might, then, be better envisaged as a domain of entanglement. It is within such a tangle of interlaced trails, continually ravelling here and unravelling there, that beings grow or "issue forth" along the lines of their relationships. This tangle is the texture of the world. In the animic ontology, beings do not simply occupy the world, they *inhabit* it, and in so doing—in threading their own paths through the meshwork—they contribute to its ever-evolving weave.
>
> (2010, p. 71, emphasis ours)

"Speaking of trails, where am I now?" you suddenly ask yourself as your reflection comes to an abrupt end. As there are no maps of the area, all you have available to retrace your steps are your memory and the traces of your footsteps. Mushroom picking, after all, is not done alongside well-established paths; it unfolds off-trail, deep into the bush. After a few sure steps back toward the direction you came from, you suddenly feel lost. No reason to panic, of course. You know it is just about noon and since you live in the northern hemisphere the sun should point you south. You know you came roughly from the west so you start walking away by keeping the sun to your left. After a few minutes you find the deer trail that you had followed earlier in the morning. "A deer trail," you think, "what an interesting idea."

In moving around the forest the deer, just like we humans do, weave their own paths. Like you right now, deer have no maps. Like you they weave into the world lines of their movement, lines of growth which occur along paths of their own making. Remember our earlier discussion of the fish? Deer, and you, are the same way: they are their movements, the trails they have woven, and the patterns of their activity. The same goes for the mushrooms you have picked and the little one you did not pick: their being is their growth, their movement through the ground and the air. And the same goes for the sun too: the line of its movement is what you knotted the line of your own movement with. You turn to Ingold and the meshwork once again: "wherever there is life there is movement," he writes. He goes on (Ingold, 2010, p. 72):

> Every creature, as it "issues forth" and trails behind, moves its characteristic way. The sun is alive because of the way it moves through the firmament, but so are the trees because of the particular ways their boughs sway or their leaves flutter in the wind, and because of the sounds they make in doing so.

You are really starting to understand how wilderness is a meshwork. A look at your own self might reinforce this point too. Your hair is full of tiny little twigs, pieces of dried-up leaves, and a couple of itsy bitsy bugs. Your trousers are dirty at the knees and wet at the bottom. Your boots are muddy. Your sweatshirt smells like last night's campfire. Traces of your movement *with* (not so much *across*) the forest pervade your body. "How different this would be," you ruminate, "if I was walking through the aisles of a supermarket looking for portobello mushrooms!" Clean, fair-smelling, in your spiffy city clothes, you might perhaps kneel to the floor of the grocery store to pick up the brown paper bag that just slipped out of your fingers and your trousers would come up unsullied from their contact with the surface. Encased by the building you would be fully protected from weather, while, if you were in some parts of North America, the mushrooms and all the other vegetables on the stand would receive a pre-programmed misty shower, perhaps accompanied by a fake thunder sound every 15 minutes. "I understand now, wilderness is a meshwork, whereas the built and developed isn't!" you would conclude.

You could not be farther from the truth. Take that very grocery shop. The roads that stretch out of its car park connect it to motorways, airways, and shipping waterways that connect it to fields and commodity chains as remote as Ecuador, Spain, New Zealand, and Chile. The electrical wires that extend to the nearest pole link it to the hydroelectrical power dam and to the river up the mountains. The trails of spoiled produce and waste entangle it with the local landfill and compost-processing plant. The scanners at the till connect with credit card networks that beep ones and zeros all the way to the financial capitals of the world. That shiny grocery shop with its cardboard-tasting farmed mushrooms is as much of a meshwork as the forest is. What now?

Instead of envisioning wilderness as the opposite of civilization, instead of imagining pristine nature as the counterpart of culture, society, or technonature, the metaphor of the meshwork invites us to think of all these things not so much as nouns which derive their significance from antonyms but rather as verbs which derive their unique meaningfulness from their own doing, their unfolding, their becoming. Let us return to our mushroom patch in the forest to further illustrate this point. Earlier we said that you were so excited at the sight of the boletes because you had not seen any in a while. They hadn't been in season, we pointed out. So, when and where do they grow exactly? Boletes' preferred habitat is a forest area dominated by pine, hemlock, spruce, and fir. Concentrated in cool temperate regions, in Europe boletes grow from Scandinavia to Italy, and in North America from Quebec to Mexico. In Asia they are found in Heilongjiang province and in the Yunnan Guizhou Plateau in China, as well as in Nepal and Tibet. In all of these and a few other regions of the world boletes spring up generally in summer and autumn, typically triggered by intermittent rainfalls during warm periods of weather. (Italian folklore even has it that a new moon is also particularly conducive to their growth!) High air humidity is also known to positively affect their growth, whereas droughts, limited soil humidity, cold night-time temperatures, and frost are widely believed to kill their chances at life. Bolete mushrooms (like other forms of life) inhabit a world of weather: that is, a weatherworld with whose activities they are deeply entangled. Ingold (2010, p. 73, original emphases) observes:

> Living beings . . . make their way *through* a nascent world rather than *across* its preformed surface. As they do so, and depending on the circumstances, they may experience wind and rain, sunshine and mist, frost and snow, and a host of other weather-related phenomena, all of which fundamentally affect their moods and motivations, their movements and their possibilities of subsistence, even as these phenomena sculpt and erode the plethora of surfaces upon which inhabitants tread. For them the inhabited world is constituted in the first places by the aerial flux of weather rather than by the grounded fixities of landscape. The weather is dynamic, always unfolding, ever changing in its currents, qualities of light and shade, and colours, alternately damp or dry, warm or cold, and so on.

Inhabiting a weatherworld, following Ingold's ideas, stands in opposition to the act of merely *occupying* a surface pre-made for life. The wildness of the bolete is embodied in its very intricate relation with the ground and the air it inhabits. Wilderness as a place, for the bolete, thus becomes a medium for growth, a place where the interlinking of earth and sky makes—or alternately prevents—movement and becoming. The same could be said for the deer whose trail you followed in and out of the forest. The deer too is entangled in the weatherworld, constantly searching for shade to rest under, for fresh grass to eat, and for a source of water to drink from as the changing seasons bring new creeks to life or dry out old ones. Mushrooms, deer, humans are not so much beings but becomings, continuously aligning their doings, their movements, "in counterpoint to the modulations of day and night, sunlight and shade, wind and weather" (Ingold, 2010, p. 88). From all of this there follows one clear and utterly important point for our understanding of wilderness.

Wilderness is a place made and constantly remade by and through the unfolding of wildness. Nevertheless, the dominant logic of inversion tricks us into thinking of wilderness as the opposite: a place that features qualities we have classified as wild according to an abstract conceptualization of that noun. But if we abandon the logic of inversion we realize that wilderness is something that comes not with the absence of human occupation of a territory. Rather, it is something occurring with the entanglements of a myriad of movements and myriad becomings. "To find a way forward," Ingold (2010, p. 114) writes in relation to our limited comprehension of life "on" earth, "we have to recognise that our humanity is neither something that comes with the territory, with our species-specific nature, nor an imagined condition that places the territory outside ourselves, but rather the ongoing historical process of our mutual and collective self-creation." Ingold's words refer to our comprehension of earthly life but they are clearly illuminating in the case of wilderness life as well. By imagining earthly life as a meshwork, and as therefore wilderness as well, we can envision the shaping of the medium we inhabit as the inescapable condition of inhabitation. To say this in other words: by inhabiting the world through our multiple relations, we shape one another and thus the world itself. The shape of any place, the shape of wilderness, "emerges, whether in the imagination or on the ground, or both simultaneously, through our very practices of inhabitation" (Ingold, 2010, p. 114). Ingold's (2010, pp 80–81) understanding of inhabitation draws significantly from writings by Jakob von Uexküll who introduced the concept of *Umwelt*. As Ingold (2010) states, "the life of every creature, von Uexküll thought, is so wrapped up in its own Umwelt that no other worlds are accessible to it. It is as though each one were floating in its own 'bubble' of reality" (p. 80).

From all of the above it follows that wilderness is wilding and rewilding, constantly growing, sprouting, and adapting as a meshwork of material flows, embodied doings, experiences, narratives, representations, and thus continuously changing as a result of the emerging entanglements of all of these with its myriad inhabitants. Wilderness, we find, is a wonderful example of a life lived "in the open" (Ingold, 2010, pp. 115–125): a life lived not between binary oppositions of being and becoming, society and wilderness, culture and nature, earth and sky, and a world of mutually exclusive hemispheres, but rather a life lived in an emerging weatherworld. A weatherworld in which every inhabitant has to contend with rain, wind, drought, snow, cold and heat as a basis for the continuation of our existence.

To experience wilderness is to be bound with the incessant movements of other emergent becomings, to come to terms with the unpredictable and unconstrained doings of the forces around us, to be fully immersed in the fluxes of the medium and the substances that become intermingled in both habitual and novel ways every hour and every day. Understood as life lived in the open, wilderness turns into something other than a neatly bounded space that we must keep distant from and observe detachedly as unwelcome visitors. Understood this way—as we cease thinking of ourselves as observers who do not remain in, who do not mess with, who do not trammel or inhabit the wild—we begin to imagine ourselves as *participants* in the wild: "each immersed with the whole of our being in the currents of a world-in-formation, in the sunlight we see in, the rain we hear in and the wind we feel in" (Ingold, 2010, p. 129).

WAYFINDING A NEW PATH THROUGH THE WILD

By understanding wilderness as an assemblage we have gained a new appreciation of the more-than-human relations that bind it together. Then, by reimagining wilderness as a meshwork we have begun to explore how wilderness is a verb, not a noun. Wilderness as meshwork is the binding of its inhabitants acting, moving, and doing what they do as they come to terms with the weatherworld, as they live life in the open. We have come a long way in this chapter and in this book in our attempt to understand how wilderness is done. We have explored how wilderness is both a real place and an abstract ideal, how places and ideals have changed throughout recent history, and how in the process of changing they have informed one another. By focusing on experiences, practices, and representations of wilderness we have examined how different wilderness areas are entangled in many different ways across the world. And by dissecting the politics of wilderness we have analyzed how conservation policies and development schemes have often turned wilderness places into contested sites. It is now time to depart, and given our pragmatic approach to wilderness we thought we would end not with a conclusive answer as to what wilderness is, but rather with something much more open-ended, something meant to allow for more doings, more actions, and more reimaginations.

If we agree that wilderness is an entanglement and that therefore our job as interpreters of this unique entity is to disentangle it, and then perhaps to reassemble it in novel and more imaginative ways, then we need a few practical tools to apply to our task. These kinds of tools should be adaptable to a multitude of different jobs, diverse purposes, differing contexts, and evolving circumstances. So, we do not want to end up proposing a set scale for measuring wilderness or for ascertaining its character. These benchmark-type tools are invariably shaped by the geo-cultural and political situations they are created in, so they are inevitably partial in their, say, Americanness, Canadianness, Australianness, South Africanness, and so on. We need tools that anyone can use anywhere, and for that matter we need to understand that anyone, anywhere, any time can simply add tools to the toolbox we get started here. The best tools—we believe—for this kind of job are interrogatives; questions that allow us to determine how wilderness is done, rather than to determine whether something is enough of a wilderness or not. In this sense we wilderness scholars should look to our urban studies colleagues for inspiration. There are, we would guess, about 500 studies into the urban condition for every single study of wilderness. And why is that? Because more people live in cities than they do in wilderness areas? Perhaps, but more possibly because for the longest time wilderness studies, unlike urban studies, have uselessly obsessed over the nature of their subject matter. How often do you find studies that obsess over the nature of what is a city? Not often, right? That is because their questions are less concerned with the essential nature of their subject and more with its unfolding and becoming, with what their subjects actually do. So, if we are to agree that wilderness is what wilderness does, then what does wilderness actually do? What are the verbs—not the nouns or qualities, but rather the actions—that allow us to understand how wilderness places become enmeshed in the unique ways they do? In closing our book, we depart with six such verbs and six final questions.

First, how does wilderness *change* and how has it changed? As we saw in the introduction to this book, wilderness has undergone, and continuously undergoes, reassembling. Wilderness is defined, understood, and conceptualized differently across historical periods and different societies. Wilderness does not remain the same. Wilderness appears where previously there was something else. Wilderness disappears, replaced by different places and competing spatializations. Our task as wilderness scholars is thus to ensure we comprehend wilderness as it mutates, evolves, adapts, and changes, rather than to determine whether or not it can or should stay the same.

Second, in what rhythms does it *grow*? As we saw in Chapter Two wilderness has been around for a very long time and has been changing for a very long time. Yet, the rhythms and speeds at which it has changed have been all but constant. Certain political, economic, social, and historical events have either accelerated or slowed down the growth of wilderness, both as an idea and as a geographical space. Tracing how the growth of wilderness has occurred in parallel with changing entanglements means better understanding its meaningfulness, its place in society, and its biological rhythms.

Third, how is wilderness *imagined*? Is it envisioned as a place of terror and lack of morality? Is it feared as the end of civilization? Is it romanticized as the epitome of the sublime? Is it rendered as a place of enchantment and wonder? In Chapter Three we examined many different ways in which our wilderness imagination is framed and reframed by competing and sometimes coexisting knowledges and stories. Future research should continue to address how various media and various communication processes represent, display, sell, and reassemble wilderness and how these assembling practices change as broader social and environmental forces confront us to reimagine the notion of the wild.

Fourth, how is it *engaged with* and *experienced*? Is wilderness envisioned as a controlled environment where interaction is predictable and orderly? Or is it understood as a place where life is lived in the open, fully unmeshed in the uncontrolled patterns of the weatherworld and the undomesticated ways of the animal world? Of course these two questions point to extreme ends of a continuum, a continuum characterized by balance of order and disorder, predictability and lack of foresight. As we saw in Chapter Four, visitors to and dwellers in wilderness turn to wild places for their transformational potential, for their ability to regenerate, to teach, and to afford experimentation through a symbolic and material break with social organization. But what are the precise ways in which these experiences and practices of wilderness unfold, and thus transform wilderness itself?

Fifth, how is wilderness inhibited, controlled, managed, or alternatively how is it let free to *become* what it wills to become? There are no right answers to this or any of the questions above. There are only questions and more questions. In Chapter Five we saw how the environmental politics surrounding the science of conservation and preservation has affected wilderness regulation and reshaped environmental awareness worldwide. More studies should examine how wilderness spaces are fought over in law and policy-making arenas, and how competing discourses, social movements, and socioeconomic forces clash over the definition of wilderness areas, thus reassembling wilderness and wildness. In surveying how wilderness is managed and administered in various ways across the world—from nature reserves, parks, and managed and unmanaged forests, to other types of protected lands and marine zones—we have reflected on how policies have impacted wilderness areas.

Finally, how is it *used*? How is wilderness divided, zoned, and controlled so as to allow certain kinds of wildness and impede others? How is wilderness fenced and classified? How is it economically categorized and according to whose priorities and values? For example, in Chapter Six we explored how many of the practices unfolding in wilderness areas and how various experiences of the wild can easily cross over into exploitations of wilderness areas for human pleasure and material gain. Tourist flows, natural resource exploitation, and business and government pressures for development and exploitation make use or attempt to make use of wilderness areas across the world in different ways, each with unique implications for wilderness. How does venturing into and out of the wilderness—we ask—leave footprints of our participation in its constant transformation of itself?

Many more questions could be asked of wilderness, wildness, and the wild. It is up to each of you to reflect on the nature of all these questions and to attempt to generate answers based on the specifics of each case. It is up to empirical research to improve our understanding of wilderness and to generate new and more useful concepts for its comprehension. For our part, our introductory task is done, hopefully, and so it is high time for us to return to our beloved forest and become immersed in its wildness and life in the open once again.

SUMMARY OF KEY POINTS

- Wilderness is an assemblage. Wilderness is something that is "thrown together" by a multitude of actors.
- Assemblages can be formally defined as "wholes characterized by relations of exteriority" (DeLanda, 2006, p. 10).
- In "relations of exteriority" we have referred to the fact that "component parts of a whole cannot be reduced to their function within that whole, and indeed they can be parts of multiple wholes at any given moment" (Dittmer, 2014, p. 387). The functions and capacities of all these various parts are deeply shaped by their finite properties and their infinite relations with the whole. Assemblages, therefore, can be understood as complex interaction systems.

- In order to disentangle assemblages, we need to focus our attention on three things: (1) the relationships between humans and non-humans; (2) the nature, directions, and rhythms of change; (3) the spatial dimensions of an assemblage.
- An ideal metaphor for understanding wilderness relationally is that of the meshwork.
- Meshworks are entanglements, knots.
- The meshwork metaphor invites us to put the logic of inversion into reverse. Meshworks force us to confront life as becoming, as movement, as something entangled in multiple currents of formation.
- Meshworks invite us to treat life as lived in the open, in a world that is not pre-constituted or occupied by things existing independent of other things, but rather whose ongoing transformation requires our inhabitation and participation.

FURTHER READINGS

DeLanda, M. (2006). *A new philosophy of society*. New York: Bloomsbury.
Ingold, T. (2010). *Being alive*. London: Routledge.
Ingold, T. (2015). *The life of lines*. London: Routledge.

References

Abram, D. (1996). *The spell of the sensuous: Perception and language in a more-than-human world.* New York: Vintage.

Abram, S. & M. E. Lien (2011). Performing nature at world's ends. *Ethnos*, 76, 3–18.

Abramson, A. & R. Fletcher (2007). Rock-climbing as epic and deep eco-play. *Anthropology Today*, 23, 3–7.

Agrawal, A. & C. Gibson (eds) (2001). *Communities and the environment: Ethnicity, gender, and the state in community-based conservation.* New Brunswick, NJ: Rutgers University Press.

Allin, C. W. (2008). *The politics of wilderness preservation.* Fairbanks, AK: Juno University of Alaska Press.

Alphandery, P. & A. Fortier (2007). A new approach to wildlife management in France: Regional guidelines as tools for the conservation of biodiversity. *Sociologia Ruralis*, 47(1), 42–62.

Anderson, D. & R. Grove (1987). *Conservation in Africa: People, policies and practice.* Cambridge, MA: Cambridge University Press.

Anderson, J. (2009). Transient convergence and relational sensibility: Beyond the modern constitution of nature. *Emotion, Space, and Society*, 2(2), 120–127.

Anderson, J. (2013). Relational places: The surfed wave as assemblage and convergence. *Environment & Planning D*, 30, 570–587.

Anderson, K. (1995). Culture and nature at the Adelaide Zoo: At the frontiers of "human" geography. *Transactions of the Institute of British Geographers*, N.S. 20, 275–294.

Aplin, G. (2004). Kakadu National Park World Heritage Site: Deconstructing the debate, 1997–2003. *Australian Geographical Studies*, 42(2), 152–174.

Armitage, D. R., R. Plummer, F. Berkes, R. I. Arthur, A. T. Charles, I. J. Davidson-Hunt, & E. K. Wollenberg (2008). Adaptive co-management for social-ecological complexity. *Frontiers in Ecology and the Environment*, 7(2), 95–102.

Aronczyk, M. (2005). Taking the SUV to a place it's never been before: SUV ads and the consumption of nature. *Invisible Culture: An Electronic Journal for Visual Culture*, 9, 1–15.

Arts, K., A. Fischer, & R. Van der Wal (2012). The promise of wilderness between paradise and hell: A cultural-historical exploration of a Dutch national park. *Landscape Research*, 37(3), 239–256.

Bagust, P. (2008). "Screen natures": Special effects and edutainment in "new" hybrid wildlife documentary. *Journal of Media & Cultural Studies*, 22(2), 213–226.

Baker, J. (2002). Production and consumption of wilderness in Algonquin Park. *Space & Culture*, 5, 198–210.

Baldwin, A. (2009). The white geography of Lawren Stewart Harris: Whiteness and the performative coupling of wilderness and multi-culturalism in Canada. *Environment & Planning A*, 41, 529–544.

Baldwin, A. (2010). Wilderness and tolerance in Flora MacDonald Denison: Towards a biopolitics of whiteness. *Social & Cultural Geography*, 11, 883–901.

Balint, P. J. & J. Mashinya (2006). The decline of a model community-based conservation project: Governance, capacity, and devolution in Mahenye, Zimbabwe. *Geoforum*, 37, 805–815.

Barrett, C. B., E. H. Bulte, P. Ferraro, & S. Wunder (2013). Economic instruments for nature conservation. *Key Topics in Conservation Biology*, 2, 59–73.

Beder, S. (2001). Anti-environmentalism. In J. Barry & E. Gene (eds), *International encyclopedia of environmental politics* (pp. 34–39). New York: Routledge.

Behan, J. R., M. T. Richards, & M. E. Lee (2001). Effects of tour jeeps in a wildland setting on non-motorized recreationist benefits. *Journal of Park and Recreation Administration*, 19, 1–19.

Bell, B. J., M. R. Holmes, B. Vigneault, & B. Williams (2008). Student involvement: Critical concerns of outdoor orientation programs. *Journal of Experiential Education*, 30, 253–257.

Bell, B., M. R. Holmes, & B. Williams (2010). A census of outdoor orientation programs at four-year colleges in the United States. *Journal of Experiential Education*, 33, 1–18.

Berman, B. & R. Lanza (2010). *Biocentrism: How life and consciousness are the keys to understanding the true nature of the universe*. Dallas, TX: BenBella Books.

Berman, D. S. & J. Davis-Berman (1991). Wilderness therapy and adolescent mental health: Administrative and clinical issues. *Administration and Policy in Mental Health*, 18, 373–379.

Berry, M. & C. Hodgson (eds) (2011). *Adventure education: An introduction*. London, UK: Routledge.

Berry, W. (2010). *What are people for? Essays*. Berkeley, CA: Counterpoint.

Bettman, J., K. Russell, & K. Parry (2013). How substance abuse recovery skills, readiness to change, and symptom reduction impact change processes in wilderness therapy participants. *Journal of Child and Family Studies*, 22, 1039–1050.

Blaikie, P. (2006). Is small really beautiful? Community-based natural resource management in Malawi and Botswana. *World Development*, 34(11), 1943–1957.

Bobilya, A., L. Akey, & D. Mitchell Jr. (2011). Outcomes of a spiritually focused wilderness orientation program. *Journal of Experiential Education*, 33, 301–323.

Bogle, R. (2014). *Biocentrism dead?: Understanding the universe and nature*. USA: CreateSpace Independent Publishing Platform.

Bogost, I. (2012). *Alien phenomenology, or what it's like to be a thing*. Minneapolis: University of Minnesota Press.

Bonta, M. (2005). Becoming-forest, becoming-local: Transformations of a protected area in Honduras. *Geoforum*, 36, 95–112.

Bousé, D. (2003). False intimacy: Close-ups and viewer involvement in wildlife films. *Visual Studies*, 18(2), 123–132.

Braasch, G. (2007). *Earth under fire: How global warming is changing the world*. Berkeley, CA: University of California Press.

Brandin, E. (2009). Versions of "wild" and the importance of fences in Swedish wildlife tourism involving moose. *Current Issues in Tourism*, 12(5–6), 413–427.

Brandth, B. & M. Haugen (2005). Doing rural masculinity: From logging to outfield tourism. *Journal of Gender Studies*, 14, 13–22.

Braun, B. (1997). Buried epistemologies: The politics of nature in (post)colonial British Columbia. *Annals of the Association of American Geographers*, 87, 3–31.

Braun, B. (2002). *The intemperate rainforest: Nature, culture, and power on Canada's west coast*. Minneapolis: University of Minnesota Press.

Braun, B. (2006). Environmental issues: Global natures in the space of assemblage. *Progress in Human Geography*, 30, 644–654.

Braun, B. & N. Castree (eds) (1998). *Remaking reality: Nature at the millennium*. London: Routledge.

Brayley, R. E. & K. M. Fox (1998). Introspection and spirituality in the backcountry recreation experience. In M. D. Bialeschki & W. P. Stewart (eds), *Abstracts from the 1998 Symposium on Leisure Research* (p. 24). Ashburn, VA: National Recreation and Parks Association.

Brereton, D. P. (2010). *Campsteading: Family, place, and experience at Squam Lake, New Hampshire*. New York: Routledge.

Brockington, D. (2002). *Fortress conservation: The preservation of the Mkomazi Game Reserve, Tanzania*. Oxford, UK: International African Institute.

Brown, C., R. McMorran, & M. F. Price (2011). Rewilding – A new paradigm for nature conservation in Scotland? *Scottish Geographical Journal*, 127(4), 288–314.

Bryant, L. R. (2010). Ken Burns and American mythology in *The national parks: America's best idea*. *Environmental Communication*, 4(4), 475–483.

Bryant, L. R. (2011). *The democracy of objects*. Ann Arbor, MI: Open Humanities Press.

Bryant, R. L. & L. Jarosz (2004). Editorial: Thinking about ethics in political ecology. *Political Geography*, 23, 807–812.

Buckley, R. (2003). Adventure tourism and the clothing, fashion and entertainment industry. *Journal of Ecotourism*, 2(2), 126–134.

Bulbeck, C. (2005). *Facing the wild: Ecotourism, conservation and animal encounters*. London: Earthscan.

Burke, S., N. Durand-Bush, & K. Doell (2010). Exploring feel and motivation with recreational and elite Mt. Everest climbers: An ethnographic study. *International Journal of Sport & Exercise Psychology*, 8, 373–393.

Burnham, P. (2000). *Indian country, God's country: Native Americans and the national parks*. Washington, DC: Island Press.

Butler, R. W. & S. W. Boyd (2000). *Tourism and national parks: Issues and implications*. Chichester, UK: John Wiley & Sons.

Bye, L. (2003). Masculinity and rurality at play in stories about hunting. *Norwegian Journal of Geography*, 57, 145–153.

Byers, A. (2005). Contemporary human impacts on alpine ecosystems in the Sagarmatha (Mt. Everest) National Park, Khumbu, Nepal. *Annals of the Association of American Geographers*, 95(1), 112–140.

Cachelin, A., J. Rose, D. Dustin, & W. Shooter (2012). Sustainability in outdoor education: Rethinking root metaphors. *Journal of Sustainability Education, 2*. Available at: www.jsedimensions.org/wordpress/content/sustainability-in-outdoor-education-rethinking-root-metaphor_2011_03/ (last accessed December 2, 2015).

Cagalanan, D. (2013). Integrated conservation and development: Impacts on households in a Philippine park. *Journal of Environment and Development*, 22(4), 435–458.

Callicott, J. B. (2008). Contemporary criticisms of the received wilderness idea. In M. Nelson & J. B. Callicott (eds), *The wilderness debate rages on* (pp. 355–377). Athens, GA: University of Georgia Press.

Callicott, J. B. & J. McRae (eds) (2014). *Environmental philosophy in Asian traditions of thought*. Albany: State University of New York Press.

Callicott, J. B. & M. P. Nelson (eds) (1998). *The great new wilderness debate*. Athens, GA: University of Georgia Press.

Carr, E. (2007). *Mission 66: Modernism and the national park dilemma*. Boston: University of Massachusetts Press.

Carr, E. (2009). *Wilderness by design: Landscape architecture and the National Park Service*. Omaha: University of Nebraska Press.

Carrier, J. G. & D. V. L. MacLeod (2005). Bursting the bubble: The socio-cultural context of ecotourism. *Journal of the Royal Anthropological Institute*, 11(2), 315–334.

Carroll, C. (2014). Native enclosures: Tribal national parks and the progressive politics of environmental stewardship in Indian Country. *Geoforum*, 53, 31–40.

Carver, S. (1998). Mapping the wilderness continuum. *International Journal of Wilderness*, 4, 1–12.

Carver, S., A. Evans, & S. Fritz (2002). Wilderness attribute mapping in the United Kingdom. *International Journal of Wilderness*, 8, 24–29.

Carver, S., J. Tricker & P. Landres (2013). Keeping it wild: Mapping wilderness character in the United States. *Journal of Environmental Management*, 131, 239–255.

Castree, N. (2005). *Nature*. London: Routledge.

Castree, N. & C. Nash (2006). Posthuman geographies. *Social and Cultural Geography*, 7(4), 501–504.

Catton, T. (1997). *Inhabited wilderness: Indians, Eskimos, and national parks in Alaska*. Santa Fe: University of New Mexico Press.

Chapin, M. (2004). A challenge to conservationists. *World Watch*, November/December, 17–31.

Chapin III, F. S., G. P. Kofinas, & C. Folke (2009). *Principles of ecosystem stewardship: Resilience-based natural resource management in a changing world*. Dordrecht, Netherlands: Springer.

Chief Luther Standing Bear (1998). Indian wisdom. In J. B. Callicott & M. P. Nelson (eds), *The great new wilderness debate* (pp. 201–206). Athens, GA: University of Georgia Press.

Clapp, R. A. (2004). Wilderness ethics and political ecology: Remapping the Great Bear Rainforest. *Political Geography*, 23, 839–862.

Cloke, P. & H. Perkins (2002). Commodification and adventure in New Zealand tourism. *Current Issues in Tourism*, 5, 521–549.

Cloke, P. & H. C. Perkins (2005). Cetacean performance and tourism in Kaikoura, New Zealand. *Environment and Planning D: Society and Space*, 23, 903–924.

Coble, T., S. Selin, & B. Erickson (2003). Hiking alone: Understanding fear, negotiation strategies, and leisure experience. *Journal of Leisure Research*, 35, 1–22.

Coggins, C. R. (2000). Wildlife conservation and bamboo management in china's southwest uplands. *Geographical Review*, 90(1), 83–111.

Cole, D. N. & D. R. Williams (2012). Wilderness visitor experiences: Lessons from 50 years of research. In D. N. Cole (ed.), *Wilderness visitor experiences: Progress in research and management; 2011 April 4–7; Missoula, MT*. (RMRS-P-66). Fort Collins, CO: US Department of Agriculture, Forest Service, Rocky Mountain Research Station.

Cole, D. N. & L. Yung (eds) (2010). *Beyond naturalness: Rethinking park and wilderness stewardship in an era of rapid change*. Washington, DC: Island Press.

Cowley, J., P. Landres, M. Memory, D. Scott, & A. Lindholm (2011–2012). Integrating cultural resources and wilderness character. *Park Science*, 28, 29–33; 38.

Cox, R. (2013). *Environmental communication and the public sphere*. Thousand Oaks, CA: SAGE.

Craig, D., L. Yung, & W. Borrie (2012). "Blackfeet belong to the mountains": Hope, loss, and Blackfeet claims to Glacier National Park, Montana. *Conservation and Society*, 10(3), 232–242.

Cronon, W. (1983). *Changes in the land: Indians, colonists, and the ecology of New England*. New York: Hill and Wang.

Cronon, W. (ed.) (1996a). *Uncommon ground: Rethinking the human place in nature*. New York: W. W. Norton & Co.

Cronon, W. (1996b). The trouble with wilderness; or getting back to the wrong nature. In W. Cronon (ed.), *Uncommon ground: Rethinking the human place in nature* (pp. 69–113). New York: W. W. Norton & Co.

Cronon, W. (2008). The riddle of the Apostle Islands: How do you manage a wilderness full of human stories? In M. Nelson & J. B. Callicott (eds), *The wilderness debate rages on* (pp. 632–644). Athens, GA: University of Georgia Press.

Crooks, K. R. & M. E. Soule (1999). Mesopredator release and avifaunal extinctions in a fragmented system. *Nature*, 400, 563–566.

Crouch, D. (2000). Places around us: Embodied lay geographies in leisure and tourism. *Leisure Studies*, 19, 63–76.
Cruikshank, J. (2006). *Do glaciers listen? Local knowledge, colonial encounters, and social imagination*. Vancouver: UBC Press.
Csikszentmihalyi, M. (1975). *Beyond boredom and anxiety*. San Francisco: Jossey-Bass.
Csikszentmihalyi, M. (1990). *Flow: The psychology of optimal experience*. New York: Harper & Row.
Csikszentmihalyi, M. & I. S. Csikszentmihalyi (1999). Adventure and the flow experience. In J. Miles & S. Priest (eds), *Adventure programming* (pp. 149–157) State College, PA: Venture.
Curtin, S. (2009). Wildlife tourism: The intangible, psychological benefits of human–wildlife encounters. *Current Issues in Tourism*, 12, 451–474.
Curtin, S. C. (2006). Swimming with dolphins: A phenomenological exploration of tourist recollections. *International Journal of Tourism Research*, 8, 301–315.
Curtin, S. C. & K. Wilkes (2005). British wildlife tourism operators: Current issues and typologies. *Current Issues in Tourism*, 8, 455–478.
Danby, R. K., D. S. Hik, D. S. Slocombe, & A. Williams (2003). Science and the St. Elias: An evolving framework for sustainability in North America's highest mountains. *Geographical Journal*, 169(3), 191–204.
Daniel, B. (2007). Life significance of a spiritually oriented, Outward Bound-type wilderness expedition. *Journal of Experiential Education*, 29, 386–390.
Darkey, D. & P. Alexander (2014). Trophy hunting: The Tuli safari circle in Zimbabwe. *Journal of Human Ecology*, 45(3), 257–268.
Davidov, V. (2012). Saving nature or performing sovereignty? Ecuador's initiative to keep oil in the ground. *Anthropology Today*, 28(3), 12–15.
Day, M. (2007). The karstlands of Antigua, their land use and conservation. *Geographical Journal*, 173(2), 170–185.
Dawson, C. & J. Hendee (2008). *Wilderness management*. Fourth edition. New York: Fulcrum Press.
Dean, B. P. (2007). Natural history, romanticism, and Thoreau. In M. Lewis (ed.), *American Wilderness: A New History* (pp. 73–89). New York: Oxford University Press.
DeLanda, M. (2006). *A new philosophy of society: Assemblage theory and social complexity*. New York: Continuum.
Deleuze, G. & F. Guattari (1987). *A thousand plateaus: Capitalism and schizophrenia*. Minneapolis: University of Minnesota Press.
DeLuca, K. & A. Demo (2008). Imagining nature and erasing class and race: Carleton Watkins, John Muir and the construction of wilderness. In M. Nelson & J. B. Callicott (eds), *The wilderness debate rages on* (pp. 189–217). Athens, GA: University of Georgia Press.
Dempsey, J. (2011). The politics of nature in British Columbia's Great Bear Rainforest. *Geoforum*, 42, 211–221.
Denevan, W. (1992). The pristine myth: The landscape of the Americas in 1492. *Annals of the Association of American Geographers*, 82, 369–385.
Desmond, C. (1999). *Staging tourism*. Chicago: University of Chicago Press.

Devall, B. & G. Sessions (2001). *Deep ecology: Living as if nature mattered.* Salt Lake City, UT: Gibbs Smith.

Dittmer, J. (2014). Geopolitical assemblages and complexity. *Progress in Human Geography*, 38, 385–401.

Dizard, J. (1994). *Going wild: Hunting, animal rights, and the contested meaning of nature.* Boston: University of Massachusetts Press.

Dizard, J. E. (2003). *Mortal stakes: Hunters and hunting in contemporary America.* Amherst, MA: University of Massachusetts Press.

Donaldson, G. W. & L. E. Donaldson (1958). Outdoor education: A definition. *Journal of Health, Physical Education, and Recreation*, 29, 17–63.

Dorwart, C. E., R. L. Moore, & Y. Leung (2009). Visitors' perceptions of a trail environment and effects on experiences: A model for nature-based recreation experiences. *Leisure Sciences: An Interdisciplinary Journal*, 32, 33–54.

Dowie, M. (2006). The hidden cost of paradise. *Stanford Social Innovation Review*, Spring, 31–38.

Dowie, M. (2009). *Conservation refugees: The hundred-year conflict between global conservation and native peoples.* London, UK: MIT Press.

Dustin, D., K. Bricker, & K. Schwab (2009). People and nature: Toward an ecological model of health promotion. *Leisure Sciences*, 32, 3–14.

Eagles, P. F. J. & S. F. McCool (2004). *Tourism in national parks and protected areas: Planning and management.* Cambridge, MA: CABI.

Echeverria, J. & R. Eby (1995). *Let the people judge: Wise use and the private property rights movement.* Washington, DC: Island Press.

Eden, S. & P. Barratt (2010). Outdoor versus indoor? Angling ponds, climbing walls and changing expectations of environmental leisure. *Area*, 43, 487–493.

Edensor, T. (2000). Walking in the British countryside: Reflexivity, embodied practices, and ways to escape. *Body & Society*, 6, 81–106.

Egoh, B. N., P. J. Farrell, A. Charef, L. J. Gurney, T. Koellner, H. Niam Abi, M. Egoh, & L. Willemen (2012). An African account of ecosystem service provision: Use, threats and policy options for sustainable livelihoods. *Ecosystem Services*, 2(1), 71–81.

Elkins, D. N., L. Hedstrom, L. Hughes, J. Leaf, & C. Saunders (1988). Toward a humanistic phenomenological spirituality: Definition, description, and measurement. *Journal of Humanistic Psychology*, 28, 5–18.

Epler, B. (2007). *Tourism, the economy, population growth, and conservation in Galápagos.* Puerto Ayora, Galápagos: Charles Darwin Foundation.

Ewah, J. O. (2012). The effects of creating access roads on the integrity of conserved areas: A case study of Okwangwo Rainforest in Cross River National Park, Nigeria. *Journal of Human Ecology*, 38(2), 105–115.

Fig, D. (2008). Stripping the desert: Uranium mining inside Namibia's Namib-Naukluft National Park. *South African Review of Sociology*, 39(2), 245–261.

Fischer, H. W. & A. Chhatre (2013). Environmental citizenship, gender, and the emergence of a new conservation politics. *Geoforum*, 50, 10–19.

Fischer, R. & E. Attah (2001). City kids in the wilderness: A pilot-test of Outward Bound for foster care group home youth. *Journal of Experiential Education*, 24, 109–118.

Flad, H. (2009). The parlor in the wilderness: Domesticating an iconic American landscape. *Geographical Review*, 99, 356–376.

Flannery, T. (2005). *The weather makers: How man is changing the climate and what it means for life on earth*. New York: Grove Press.

Foreman, D. (2014). *The great conservation divide: Conservation vs. resourcism on America's public lands*. Durango, CO: Ravens Eye Press.

Foreman, D., J. Davis, D. Johns, R. Noss, & M. Soule (1992). The Wildlands Project mission statement. *Wild Earth*, 2(1), 3–4.

Fox, R. (1999). Enhancing spiritual experience in adventure programs. In J. C. Miles & S. Priest (eds), *Adventure programming* (pp. 455–461). State College, PA: Venture Publishing.

Fox, R. J. (1997). Women, nature and spirituality: A qualitative study exploring women's wilderness experience. In D. Rowe & P. Brown (eds), *Proceedings, ANZALS conference 1997* (pp. 59–64). Newcastle, NSW, Australia: Australian and New Zealand Association for Leisure Studies, and Department of Leisure and Tourism Studies, The University of Newcastle.

Franklin, A. (1999). *Animals and modern cultures: A sociology of human–animal relations in modernity*. London: SAGE.

Franklin, A. (2001). Neo-Darwinian leisures, the body, and nature: Hunting and angling in modernity. *Body & Society*, 7, 57–67.

Franklin, A. (2008). The animal question and the consumption of wildlife. In B. Lovelock (ed.), *Tourism and the consumption of wildlife: Hunting, shooting and sport fishing* (pp. 31–43). London: Routledge.

Fredrickson, L. M. & D. Anderson (1999). A qualitative exploration of the wilderness experience as a source of spiritual inspiration. *Journal of Environmental Psychology*, 19, 21–39.

Frost, W. (2004). Tourism, rainforests and worthless lands: The origins of national parks in Queensland. *Tourism Geographies*, 6(4), 493–507.

Frost, W. (2010). Life changing experiences: Films and tourists in the Australian Outback. *Annals of Tourism Research*, 37(3), 707–726.

Frost, W. & C. M. Hall (eds) (2009). *Tourism and national parks: International perspectives on development, histories and change*. London, UK: Routledge.

Gandy, M. (1996). Visions of darkness: The representation of nature in the films of Werner Herzog. *Ecumene*, 3, 1–21.

García-Amado, L. R., M. R. Pérez, & S. Barrasa García (2013). Motivation for conservation: Assessing integrated conservation and development projects and payments for environmental services in La Sepultura Biosphere Reserve, Chiapas, Mexico. *Ecological Economics*, 89, 92–100.

Gass, M. A., D. Garvey, & D. Sugerman (2003). The long-term effects of a first-year student wilderness orientation program. *Journal of Experiential Education*, 26, 34–40.

Giblett, R. (2007). Shooting the sunburnt country, the land of sweeping plains, the rugged mountain ranges: Australian landscape and wilderness photography. *Continuum: Journal of Media & Cultural Studies*, 21(3), 335–346.

Giblett, R. (2009). *Landscapes of culture and nature*. Houndmills: Palgrave Macmillan.

Giroux, H. A. (1999). *The mouse that roared: Disney and the end of innocence*. Lanham: Rowman & Littlefield.

Goffman, E. (1974). *Frame analysis*. Cambridge: Harvard University Press.

Goldman, M. (2003). Partitioned nature, privileged knowledge: Community-based conservation in Tanzania. *Development and Change*, 34(5), 833–862.

Gould, R. K. (2005). *At home in nature: Modern homesteading and spiritual practice in America*. Berkeley, CA: University of California Press.

Grafanaki, S., D. Pearson, F. Cini, B. Godula, B. McKenzie, S. Nason, & M. Anderegg (2005). Sources of renewal: A qualitative study of the experience and role of leisure in the life of counselors and psychologists. *Counselling Psychology Quarterly*, 18, 31–34.

Great Barrier Reef Marine Authority (GBRMA) (1994). *The Great Barrier Reef—keeping it great: A 25 year strategic plan for the Great Barrier Reef World Heritage Area*. Townsville: GBRMA.

Greig, J. & T. Whillains (1998). Restoring wilderness functions and the vicarious basis of ecological stewardship. *Journal of Canadian Studies*, 33, 116–125.

Grunewald, K. & O. Bastian (eds) (2015). *Ecosystem services: Concept, methods, and case studies*. New York: Springer.

Guha, R. (1998). Deep ecology revisited. In J. B. Callicott & M. P. Nelson (eds), *The great new wilderness debate* (pp. 271–279). Athens, GA: University of Georgia Press.

Haalboom, B. (2011). Framed encounters with conservation and mining development: Indigenous peoples' use of strategic framing in Suriname. *Social Movement Studies*, 10(4), 387–406.

Hackel, J. D. (1999). Community conservation and the future of Africa's wildlife. *Conservation Biology*, 13 (4), 726–734.

Hailey, C. (2008). *Campsites: Architecture of duration and place*. New Orleans: Louisiana University Press.

Hailey, C. (2009). *Camps: A guide to 21st century space*. Boston: MIT Press.

Haraway, D. (1991). *Simians, cyborgs, and women: The reinvention of nature*. New York: Routledge.

Haynes, C. (2013). Seeking control: Disentangling the difficult sociality of Kakadu National Park's joint management. *Journal of Sociology*, 49(2–3), 194–209.

Head, L. M. (2000). Renovating the landscape and packaging the penguin: Culture and nature on Summerland Peninsula, Phillip Island, Victoria, Australia. *Australian Geographical Studies*, 38(1), 36–53.

Heintzman, P. (1998). The role of introspection/spiritual in the park experience of campers at Ontario Provincial Parks. In *Culture, environment, and society* (pp. 169–170). Columbia, MS: University of Missouri-Columbia.

Heintzman, P. (2000). Leisure and spiritual well-being relationships: A qualitative study. *Society and Leisure*, 23, 41–69.

Heintzman, P. (2002). The role of introspection and spirituality in the park experience of day visitors to Ontario Provincial Parks. In S. Bondrup-Nielsen, N. Munro, G. Nelson, M. Willison, T. Herman, & P. Eagles (eds), *Managing*

protected areas in a changing world (pp. 992–1004). Wolfville, Nova Scotia, Canada: Science and Management of Protected Areas Association.
Heintzman, P. (2007a). Men's wilderness experience and spirituality: A qualitative study. In R. Burns & K. Robinson (Comps.), *Proceedings of the 2006 Northeastern Recreation Research Symposium* (pp. 432–439). Newton Square, PA: US Department of Agriculture, Forest Services, Northern Research Station.
Heintzman, P. (2007b). Rowing, sailing, reading, discussing, praying: The spiritual and lifestyle impact of an experientially based, graduate, environmental education course. Paper presented at the Trails to Sustainability Conference, Kananaskis, AB, Canada.
Heintzman, P. (2009). Nature-based recreation and spirituality: A complex relationship. *Leisure Sciences*, 32, 72–89.
Hendee, J. C. & C. P. Dawson (2002). *Wilderness management: Stewardship and protection of resources and values*. Golden, CO: Fulcrum Publishing.
Hendee, J., G. Stankey, & R. Lucas (1990). *Wilderness management*. Golden, CO: North American Press.
Hennessy, E. A. & A. L. McCleary (2011). Nature's Eden? The production and effects of "pristine" nature in the Galápagos Islands. *Island Studies Journal*, 6(2), 131–156.
Hepburn, S. J. (2013). In Patagonia (clothing): A complicated greenness. *Fashion Theory: The Journal of Dress, Body & Culture*, 17(5), 623–645.
Hermer, J. (2002). *Regulating Eden: The nature of order in North American parks*. Toronto: University of Toronto Press.
Hill, J., S. Curtin, & G. Gough (2014). Understanding tourist encounters with nature: A thematic framework. *Tourism Geographies*, 16, 68–87.
Hill, K. (2013). Wayfinding and spatial reorientation by Nova Scotia deer hunters. *Environment & Behavior*, 45, 267–283.
Hilson, G. & F. Nymae (2006). Gold mining in Ghana's forest reserves: A report on the current debate. *Area*, 38(2), 175–185.
Himley, M. (2009). Nature conservation, rural livelihoods, and territorial control in Andean Ecuador. *Geoforum*, 40, 832–842.
Hintz, J. (2007). Some political problems for rewilding nature. *Ethics, Place and Environment: A Journal of Philosophy and Geography*, 10(2), 177–216.
Hitzhusen, G. (2004). Understanding the role of spirituality and theology in outdoor environmental education: A mixed-method characterization of 12 Christian and Jewish outdoor programs. In K. Paisley, C. J. Bunting, A. B. Young, & K. Bloom (eds), *Research in outdoor education*: Vol. 7 (pp. 39–56). Cortland, NY: Coalition for Education in the Outdoors.
Hobbs, E. (2011). Performing wilderness, performing difference: Schismogenesis in a mining dispute. *Ethnos: Journal of Anthropology*, 76, 109–129.
Hobsbawm, E. & T. Ranger (eds) (1983). *The invention of tradition*. Cambridge, UK: Cambridge University Press.
Hodgins, B. (1998). Refiguring wilderness: A personal odyssey. *Journal of Canadian Studies*, 33, 12–26.
Hodgson, D. (2001). *Once intrepid warriors: Gender, ethnicity, and the cultural politics of Maasai development*. Bloomington, IN: Indiana University Press.

Holmes, G. (2014). Defining the forest, defending the forest: Political ecology, territoriality, and resistance to a protected area in the Dominican Republic. *Geoforum*, 53, 1–10.

Holmes, G. & D. Brockington (2013). Protected areas and wellbeing. In D. Roe, M. Walpole, C. Sandbrook, & J. Elliot (eds), *Linking biodiversity conservation and poverty reduction*. London: Wiley-Blackwell.

Honey, M. (2008). *Ecotourism and sustainable development: Who owns paradise?* Washington, DC: Island Press.

Hoole, A. & F. Barkes (2009). Breaking down fences: Recoupling social–ecological systems for biodiversity conservation in Namibia. *Geoforum*, 41, 304–317.

Huijbens, E. & K. Benediktsson (2007). Practising Highland heterotopias: Automobility in the interior of Iceland. *Mobilities*, 2, 143–165.

Hunn, E., D. Johnson, P. Russell, & T. Thornton (2003). Huna Tlingit traditional environmental knowledge, conservation, and the management of a "wilderness" park. *Current Anthropology*, 44, S79–S103.

Ingold, T. (2000). *The perception of the environment*. London: Routledge.

Ingold, T. (2007). *Lines: A brief history*. London: Routledge.

Ingold, T. (2008). Bindings against boundaries: Entanglements of life in an open world. *Environment & Planning A*, 40, 1796–1810.

Ingold, T. (2010). *Being alive*. London: Routledge.

Ingold, T. (2011). Introduction. In T. Ingold (ed.), *Redrawing anthropology: Materials, movements, lines* (pp. 1–19). Farnham: Ashgate.

Ingold, T. (2015). *The life of lines*. London: Routledge.

Ingold, T. & T. Kurttila (2000). Perceiving the environment in Finnish Lapland. *Body & Society*, 6, 183–196.

International Union for Conservation of Nature (IUCN) (1980). World Conservation Strategy. https://portals.iucn.org/library/efiles/documents/WCS-004.pdf (last accessed December 2, 2015).

International Union for Conservation of Nature (IUCN) (2008). Protected areas. www.iucn.org/about/work/programmes/gpap_home/gpap_quality/gpap_pacategories (last accessed October 24, 2014).

International Union for Conservation of Nature (IUCN) (2014). Protected areas Category 1b. www.iucn.org/about/work/programmes/gpap_home/gpapquality/gpap_pacategories/gpap_category1b (last accessed December 2, 2015).

Jacques, P. & D. Ostergren (2006). The end of wilderness: Conflict and defeat of wilderness in the Grand Canyon. *Review of Policy Research*, 23, 573–587.

Jamal, T. & M. Eyre (2003). Legitimation struggles in national park spaces: The Banff Bow Valley round table. *Journal of Environment Planning and Management*, 46(3), 417–441.

James, W. (1902). *The varieties of religious experience*. London: Longman.

Jazeel, T. (2005). "Nature", nationhood and the poetics of meaning in Ruhuna (Yala) National Park, Sri Lanka. *Cultural Geographies*, 12(2), 199–227.

Johns, D. M. (1998). The relevance of deep ecology to the third world: Some preliminary comments. In J. B. Callicott & M. P. Nelson (eds),

The great new wilderness debate (pp. 246–270). Athens, GA: University of Georgia Press.

Johnson, C. Y. & J. M. Bowker (2008). African American wildland memories. In M. Nelson & J. B. Callicott, *The wilderness debate rages on: Continuing the great new wilderness debate* (pp. 325–349). Athens, GA: University of Georgia Press.

Jorgenson, D. (2015). Rethinking rewilding. *Geoforum*, 65, 482–488.

Kahn, P. & P. Hasbach (eds) (2013). *Rediscovery of the wild*. Cambridge, MA: MIT Press.

Kahn, P. H. Jr. (1999). *The human relationship with nature: Development and culture*. Cambridge, MA: MIT Press.

Kalisch, K., A. Bobilya, & B. Daniel (2011). The Outward Bound solo: A study of participants' perceptions. *Journal of Experiential Education*, 34, 1–19.

Kane, M. & H. Tucker (2004). Adventure tourism: The freedom to play with reality. *Tourist Studies*, 4, 217–234.

Kane, M. & R. Zink (2004). Package adventure tours: Markers in serious leisure careers. *Leisure Studies*, 23, 329–345.

Kaplan, S. (1995). The restorative benefits of nature: Toward an integrative framework. *Journal of Environmental Psychology*, 15, 169–182.

Kaufman, E. (1998). "Naturalizing the nation": The rise of naturalistic nationalism in the United States and Canada. *Comparative Studies in Society and History*, 40(4), 666–695.

Kearns, R. & J. Fagan (2014). Sleeping with the past? Heritage, recreation, and transition in New Zealand tramping huts. *New Zealand Geographer*, 70, 116–130.

Kelly, J. (1996). *Leisure*. Third edition. Boston: Allyn & Bacon.

Kelly, V. (2011). Women of courage: A personal account of a wilderness-based experiential group for survivors of abuse. *Journal for Specialists in Group Work*, 31, 99–111.

Keul, A. (2013). Embodied encounters between humans and gators. *Social & Cultural Geography*, 14, 930–953.

King, B. (2010). Conservation geographies in Sub-Saharan Africa: The politics of national parks, community conservation and peace parks. *Geography Compass*, 4(1), 14–27.

King, B. H. (2007). Conservation and community in the new South Africa: A case study of the Mahushe Shongwe Game Reserve. *Geoforum*, 38, 207–219.

Knight, J. (2010). The ready-to-view wild monkey: The convenience principle in Japanese wildlife tourism. *Annals of Tourism Research*, 37, 744–762.

Kohn, E. (2013). *How forests think: Toward an anthropology beyond the human*. Berkeley, CA: University of California Press.

Koole, S. & A. Van Den Berg (2005). Lost in the wilderness: Terror management, action orientation, and nature evaluation. *Journal of Personality and Social Psychology*, 6, 1014–1028.

Kopas, P. (2008). *Taking the air: Ideas and change in Canada's National Parks*. Vancouver: UBC Press.

Kormos, C. (ed.) (2008). *A handbook on international wilderness law and policy*. New York: Fulcrum Press.

Kortelainen, J. (2010). Old growth forests as objects in complex spatialities. *Area*, 42, 494–501.

Kosek, J. (2006). *Understories: The political life of forests in northern New Mexico*. Durham, NC: Duke University Press.

Krech, S. (2000). *Ecological Indian: Myths and history*. New York: Norton.

Kropp, P. (2009). Wilderness wives and dishwashing husbands: Comfort and the domestic arts of camping in America, 1880–1910. *Journal of Social History*, Fall, 5–30.

Kusler, J. A. (1991). Ecotourism and resource conservation. Presented at the 1st International Symposium on Ecotourism. Mexico City: Omnipress.

Lakoff, G. (2010). Why it matters how we frame the environment. *Environmental Communication*, 4, 70–81.

Landres P., P. Morgan, & F. Swanson (1999). Overview of the use of natural variability concepts in managing ecological systems. *Ecological Applications*, 9, 1179–1188.

Landres, P., S. Boutcher, L. Merigliano, C. Barns, D. Davis, & T. Hall (2005). *Monitoring selected conditions related to wilderness character: A national framework*. Gen. Tech. Report RMRS-GTR-151. Rocky Mountain Research Station, USDA Forest Service.

Landres, P., C. Barns, J. G. Dennis, T. Devine, P. Geissler, C. S. McCasland, L. Merigliano, J. Seastrand, & R. Swain (2008). *Keeping it wild: An interagency strategy to monitor trends in wilderness character across the National Wilderness Preservation System (RMRS-GTR-212)*. Fort Collins, CO: US Department of Agriculture, Forest Service, Rocky Mountain Research Station.

Larson, E. J. (2001). *Evolution's workshop: God and science on the Galápagos Islands*. New York: Basic Books.

Lasenby, J. (2003). Exploring episode-type spiritual experience associated with outdoor education programs. Unpublished master's thesis, University of Edinburgh, Scotland.

Latour, B. (1993). *We have never been modern*. Cambridge, MA: Harvard University Press.

Latour, B. (2004). *The politics of nature*. Cambridge, MA: Harvard University Press.

Laviolette, P. (2007). Hazardous sport? *Anthropology Today*, 23, 1–2.

Lawson, H. M. (2003). Controlling the wilderness: The work of wilderness officers. *Society & Animals*, 11(4), 329–351.

Lawson, S. R. & R. E. Manning (2002). Tradeoffs among social, resource, and management attributes of the Denali wilderness experience: A contextual approach to normative research. *Leisure Sciences*, 24, 297–312.

Lemelin, R. & E. Wiersma (2007). Gazing upon Nanuk, the polar bear: The social and visual dimensions of the wildlife gaze in Churchill, Manitoba. *Polar Geography*, 30, 37–53.

Lemelin, R. H. (2006). The gawk, the glance, and the gaze: Ocular consumption and polar bear tourism in Churchill, Manitoba, Canada. *Current Issues in Tourism*, 9(6), 516–534.

Leopold, A. (1986). *A sand county almanac: Outdoor essays and reflections*. New York: Ballantine Books.

Li, T. (2007). Practices of assemblage and community forest management. *Economy & Society*, 36, 263–293.

Lien, M. & M. Goldenberg (2012). Outcomes of a college wilderness orientation program. *Journal of Experiential Education*, 35, 253–272.

Lien, M. and J. Law (2011). "Emergent aliens": On salmon, nature, and their enactment. *Ethnos: A Journal of Anthropology*, 76, 65–87.

Lindberg, K. (1991). Policies for maximising nature tourism's ecological and economic benefits. *International Conservation Financing Project Working Paper*. Washington, DC: World Resources Institute.

Lines, W. (2006). *Patriots: Defending Australia's natural heritage*. St. Lucia, QLD: Queensland University Press.

Litchfield, C. (2001). Responsible tourism with great apes in Uganda. In S. F. McCool & R. F. Moisey (eds), *Tourism, recreation and sustainability: Linking culture and the environment* (pp. 105–32). New York: CABI.

Livengood, J. (2009). The role of leisure in the spirituality of New Paradigm Christians. *Leisure/Loisir*, 33, 389–417.

Louter, D. (2009). *Windshield wilderness: Cars, roads, and nature in Washington's national parks*. Seattle: University of Washington Press.

Lovelock, B. (ed.) (2008a). *Tourism and the consumption of wildlife: Hunting, shooting, and sport fishing*. New York: Routledge.

Lovelock, B. (2008b). Introduction. In B. Lovelock (ed.), *Tourism and the consumption of wildlife: Hunting, shooting, and sport fishing* (pp. 3–30) New York: Routledge.

Lund, K. (2005). Seeing in motion and the touching eye: Walking over Scotland's mountains. *Ethnofoor*, 18, 27–42.

Lund, K. (2008). Making mountains, producing narratives, or: "One day some poor sod will write their Ph.D. on this." *Archives & Social Studies: A Journal of Interdisciplinary Research*, 2, 135–162.

Lund, K. A. (2013). Experiencing nature in nature-based tourism. *Tourist Studies*, 13(2), 156–171.

Lunstrum, E. (2014). Green militarization: Anti-poaching efforts and the spatial contours of Kruger National Park. *Annals of the Association of American Geographers*, 104(4), 816–832.

Lunstrum, E. (2015). Green grabs, land grabs and the spatiality of displacement: Eviction from Mozambique's Limpopo National Park. *Area*, 1–11.

McCann, J. (1999a). Before 1492: The making of the pre-Columbian landscape, part 1: The environment. *Ecological Restoration*, 17, 15–30.

McCann, J. (1999b). Before 1492: The making of the pre-Columbian landscape, part 2: The vegetation, and implications for restoration for 2000 and beyond. *Ecological Restoration*, 17, 107–119.

McDermott Hughes, D. (2005). Third nature: Making space and time in the great Limpopo conservation area. *Cultural Anthropology*, 20 (2), 157–184.

McDonald, B. L. (1989). The outdoors as a setting for spiritual growth. *Women in Natural Resources*, 10, 19–23.

Mace, B. L., P. A. Bell, & R. J. Loomis (2004). Visibility and natural quiet in national parks and wilderness areas: Psychological considerations. *Environment and Behaviour*, 36(5), 5–31.

McFarlane, C. & B. Anderson (2011). Thinking with assemblage. *Area*, 43, 162–164.

MacKenzie, J. M. (1988). *The empire of nature*. Manchester, UK: Manchester University Press.

MacKenzie, J. M. (1990). *Imperialism and the natural world*. Manchester, UK: Manchester University Press.

MacKerron, G. & S. Mourato (2013). Happiness is greater in natural environments. *Global Environmental Change*, 23, 992–1000.

McLean, J. & S. Straede (2003). Conservation, relocation, and the paradigms of park and people management—A case study of Padampur Villages and the Royal Chitwan National Park, Nepal. *Society and Natural Resources*, 16, 509–526.

McLeod, A. (2014). *Understanding Asian philosophy: Ethics in the Analects, Zhuangzi, Dhammapada, and the Bhagavad Gita*. London: Bloomsbury.

Macnaghten, P. & J. Urry (1998). *Contested natures*. Thousand Oaks, CA: SAGE.

Macnaghten, P. & J. Urry (2000). Bodies in the woods. *Body & Society*, 6, 166–182.

Maddock, A. H. & M. G. L. Mills (1994). Population characteristics of African wild dogs (*lycaon pictus*) in the eastern Transvaal lowveld, South Africa, as revealed through photographic records. *Biological Conservation*, 6, 57–62.

Mann, C. (2005). *1491: New revelations of the Americas before Columbus*. New York: Knopf.

Manning, R. & L. Anderson (2012). *Managing outdoor recreation: Case studies in the national parks*. Cambridge, MA: CABI.

Marris, E. (2013). *The rambunctious garden: Saving nature in a post-wild world*. New York: Bloomsbury.

Magome, H. & J. Murombedzi (2003). Sharing South African national parks: Community land and conservation in a democratic South Africa. In W. M. Adams & M. Mulligan (eds), *Decolonizing nature: Strategies for conservation in a post-colonial era* (pp. 108–134). London: Earthscan.

Marion, J. L. & S. E. Reid (2001). Development of the US Leave No Trace programme: An historical perspective. In M. B. Usher (ed.) *Enjoyment and understanding of the natural heritage*. Scottish Natural Heritage (pp. 81–92). Edinburgh, Scotland: The Stationery Office.

Marion, J. L. & S. E. Reid (2007). Minimizing visitor impacts to protected areas: The efficacy of low impact education programmes. *Journal of Sustainable Tourism*, 15(1), 5–27.

Marks, S. (2001). Back to the future: Some unintended consequences of Zambia's community-based wildlife program (ADMADE). *Africa Today*, 48(1), 121–141.

Markus, G. & E. Saka (2006). Assemblage. *Theory, Culture & Society*, 23, 106–108.

Markwell, K. (2001). "An intimate rendezvous with nature"? Mediating the tourist–nature experience at three tourist sites in Borneo. *Tourist Studies*, 1, 39–57.

Marlowe, J., N. Pearl, & M. Marlowe (2002). School of urban wilderness survival and the circle of courage. *Reclaiming Children and Youth*, 18, 3–6.

Marsh, K. (2009). *Drawing lines in the forest: Creating wilderness areas in the Pacific Northwest*. Seattle: University of Washington Press.
Marsh, P. E. (2008). Backcountry adventure as spiritual development: A means-end study. *Journal of Experiential Education*, 30, 290–293.
Mawani, R. (2007). Legalities of nature: Law, empire, and wilderness landscapes in Canada. *Social Identities*, 13, 715–734.
Mbaiwa, J. E. (2003). The socio-economic and environmental impacts of tourism development on the Okavango Delta, North-Western Botswana. *Journal of Arid Environment*, 54(2), 447–467.
Merchant, C. (1996). *Earthcare: Women and the environment*. London, UK: Routledge.
Metcalf, E. C., A. R. Graefe, N. E. Trauntvein, & R. C. Burns (2015). Understanding hunting constraints and negotiation strategies: A typology of female hunters. *Human Dimensions of Wildlife* (20), 30–46.
Meyer-Arendt, K. (2004). Tourism and the natural environment. In A. Lew, C. Michael Hall, & A. Williams (eds), *A companion to tourism*, pp. 425–437. Oxford: Blackwell.
Michael, M. (2000). These boots are made for walking: Mundane technology, the body and human–environmental relations. *Body & Society*, 6, 126–145.
Milder, J. C., A. K. Hart, P. Dobie, & J. Minai (2014). Integrated landscape initiatives for African agriculture, development, and conservation: A region-wide assessment. *World Development*, 54(2), 68–80.
Miles, J. (2009). *Wilderness in national parks: Playground or preserve?* Seattle: University of Washington Press.
Milton, K. (1999). Nature is already sacred. *Environmental Values*, 8, 437–449.
Modelmog, I. (1998). Nature as a promise of happiness: Farmers' wives in the area of Ammerland, Germany. *Sociologia Ruralis*, 38, 109–122.
Mordue, T. (2009). Angling in modernity: A tour through society, nature and embodied passion. *Current Issues in Tourism*, 12, 529–552.
Morris, N. (2011). Night walking: Darkness and sensory perception in a night-time landscape installation. *Cultural Geographies*, 18, 315–342.
Mullins, P. (2009). Living stories of the landscape: Perception of place through canoeing in Canada's north. *Tourism Geographies*, 11, 233–255.
Mullins, P. (2014). A socio-environmental case for skill in outdoor adventure. *Journal of Experiential Education*, 37, 129–144.
Murphy, B. M. (2013). *The rural gothic in American popular culture: Backwoods horror and terror in the wilderness*. Houndmills: Palgrave Macmillan.
Murtaugh, P. A., L. Burns, & J. Schuster (1999). Predicting the retention of university students. *Research in Higher Education*, 40, 355–371.
Musa, G., C. M. Hall, & J. E. S. Higham (2004). Tourism sustainability and health impacts in high altitude adventure, cultural and ecotourism destinations: A case study of Nepal's Sagarmatha National Park. *Journal of Sustainable Tourism*, 12(4), 306–331.
Næss, A., A. Draengson, & B. Devall (2009). *The ecology of wisdom: Writings by Arne Naess*. Berkeley, CA: Counterpoint Press.

Narayan, Y. & J. MacBeth (2009). Deep in the desert: Merging the desert and the spiritual through 4WD tourism. *Tourism Geographies*, 11, 369–389.

Nash, R. F. (1982). *Wilderness and the American mind.* Third edition. New Haven, CT: Yale University Press.

Nash, R. F. (2001). *Wilderness and the American mind.* Fourth edition. New Haven, CT: Yale University Press.

National Wilderness Steering Committee (2002). Cultural resources and wilderness. Guidance "White Paper." Number 1, November. National Wilderness Steering Committee, USDI National Park Service, Washington, DC, USA.

Nelson, M. (1998). An amalgamation of wilderness preservation arguments. In J. B. Callicott & M. P. Nelson (eds), *The great new wilderness debate* (pp. 154–198). Athens, GA: University of Georgia Press.

Nelson, M. & J. B. Callicott (eds) (2008). *The wilderness debate rages on: Continuing the great new wilderness debate.* Athens, GA: University of Georgia Press.

Nepal, S. (2005). Tourism and remote mountain settlements: Spatial and temporal development of tourist infrastructure in the Mt. Everest region, Nepal. *Tourism Geographies*, 7(2), 205–227.

Neumann, M. (1999). *On the rim: Looking for the Grand Canyon.* Minneapolis: University of Minnesota Press.

Neumann, R. (1995). Ways of seeing Africa: Colonial recasting of African society and landscape in Serengeti National Park. *Cultural Geographies*, 2, 149–169.

Neumann, R. (1998). *Imposing wilderness: Struggles over livelihood and nature preservation in Africa.* Berkeley, CA: University of California Press.

Neumann, R. P. (2004). Moral and discursive geographies in the war for biodiversity in Africa. *Political Geography*, 23, 813–837.

Norton, C. & T. Watt (2013). Exploring the impact of a wilderness-based positive youth development program for urban youth. *Journal of Experiential Education*, 37, 335–350.

Noss, R. (1998). Wilderness recovery: Thinking big in restoration ecology. In J. B. Callicott & M. P. Nelson (eds), *The great new wilderness debate* (pp. 521–539). Athens, GA: University of Georgia Press.

Nustad, K. (2011). Performing natures and land in the iSimangaliso Wetland Park, South Africa. *Ethnos*, 76, 88–108.

Oelschlaeger, M. (1991). *The idea of wilderness: From prehistory to the age of ecology.* New Haven, CT: Yale University Press.

Ogden, L. (2001). *Swamplife: People, gators, and mangroves entangled in the Everglades.* Minneapolis: University of Minnesota Press.

Okech, R. N. (2010). Tourism development in Africa: Focus on poverty alleviation. *Journal of Tourism and Peace Research*, 1(1), 1–8.

O'Keefe, M. (1989). Freshman wilderness orientation programs: Model programs across the country. In J. Gilbert (ed.), *Life beyond walls: Proceedings of the National Conference on Outdoor Recreation* (pp. 165–179). Ft. Collins, CO: Colorado State University Press.

Olafsdottir, G. (2011). Practising (nature-based) tourism: An introduction. *Landabréfið*, 25, 3–14.

Olafsdottir, G. (2013a). On nature-based tourism. *Tourist Studies*, 13: 127–138.

Olafsdottir, G. (2013b). "... sometimes you've just got to get away": On trekking holidays and their therapeutic effect. *Tourist Studies*, 13(2), 209–231.

Olson, L., C. Finnegan, & D. Hope (2008). *Visual rhetoric: A reader in communication and American culture*. Thousand Oaks, CA: SAGE.

Palmer, L. (2004a). Fishing lifestyles: "Territorians", traditional owners and the management of recreational fishing in Kakadu National Park. *Australian Geographical Studies*, 42(1), 60–76.

Palmer, L. (2004b). Bushwalking in Kakadu: A study of cultural borderlands. *Social & Cultural Geography*, 5(1), 109–127.

Palmer, L. (2007). Interpreting "nature": The politics of engaging with Kakadu as an Aboriginal place. *Cultural Geographies*, 14(2), 255–273.

Palmer, P. (2006). "Nature", place and the recognition of indigenous polities. *Australian Geographer*, 37(1), 33–43.

Perreault, M. (2007). American wilderness and first contact. In M. Lewis (ed.), *American wilderness: A new history* (pp. 15–33). New York: Oxford University Press.

Phillips, J. (2006). Agencement/assemblage. *Theory, Culture & Society*, 23, 108–110.

Pierson, D. P. (2005). "Hey, they're just like us!": Representations of the animal world in the Discovery Channel's nature programming. *Journal of Popular Culture*, 38(4), 698–712.

Plummer, R. (2009). *Outdoor recreation: An introduction*. London: Routledge.

Plumwood, V. (1998). Wilderness skepticism and wilderness dualism. In J. B. Callicott & M. P. Nelson (eds), *The great new wilderness debate* (pp. 652–690). Athens, GA: University of Georgia Press.

Pollan, M. (1991). *Second nature: A gardener's education*. New York: Grove Press.

Price, J. (1999). *Flight maps: Adventures with nature in modern America*. New York: Basic Books.

Proctor, J. (1998). The social construction of nature: Relativist accusations, pragmatist and critical realist responses. *Annals of the Association of American Geographers*, 88, 352–376.

Quiroga, D. (2009). Crafting nature: The Galápagos and the making and unmaking of a natural laboratory. *Journal of Political Ecology*, 16(1), 123–140.

Rantala, O. (2010). Tourist practices in the forest. *Annals of Tourism Research*, 37, 249–264.

Reed, M. G. (2003). Marginality and gender at work in forestry communities in British Columbia, Canada. *Journal of Rural Studies*, 19, 373–389.

Riley, M. & J. Hendee (1999). Wilderness Vision Quest clients, motivations and reported benefits from an urban based program 1988–1997. In A. E. Watson, G. Aplet, & J. C. Hendee (eds), *Personal, societal and ecological values of wilderness: Sixth World Wilderness Congress proceedings on research, management, and allocation, Vol. II, Proc. RMRS-P-000* (pp. 234–249). Ogden, UT: USDA Forest Service, Rocky Mountain Research Station.

Ripple, W. J. & R. L. Beschta (2006). Linking wolves to willows via risk-sensitive foraging by ungulates in the northern Yellowstone ecosystem. *Forest Ecology and Management*, 230, 96–106.

Ritzer, G. (1993). *The McDonaldization of society*. New York: Pine Forge.

Robbins, P. (2004). *Political ecology: A critical introduction*. Malden, MA: Blackwell Publishing.

Robbins, P. & B. Marks (2009). Assemblage geographies. In S. Smith, R. Pain, S. Marston, & J. P. Jones III (eds), *The SAGE handbook of social geographies* (pp. 176–193). London: SAGE.

Roe, D. & M. Jack (eds) (2001). Stories from Eden: Studies of community-based wildlife management. In *Evaluating Eden Series*, vol. 9. London: International Institute for Environment and Development.

Rose, D. (1996). *Nourishing terrains: Australian Aboriginal views of landscape and wilderness*. Canberra: Australian Heritage Commission.

Roth, R. J. (2008). "Fixing" the forest: The spatiality of conservation conflict in Thailand. *Annals of the Association of American Geographers*, 98(2), 373–391.

Rothenberg, D. & M. Ulvaeus (2001). *The world and the wild*. Tucson, AZ: University of Arizona Press.

Rousseau, J.-J. (2004 [1782]) *The confessions of Jean-Jacques Rousseau* (E-Book #3913). Project Gutenberg. www.gutenberg.org/files/3913/3913-h/3913-h.htm (last accessed December 2, 2015).

Rugendyke, B. & N. Thi Son (2005). Conservation costs: Nature-based tourism as development at Cuc Phuong National Park, Vietnam. *Asia Pacific Viewpoint*, 46(2), 185–200.

Runte, A. (1979). *National parks: The American experience*. New York: Taylor Trade Publishing.

Russell, J. & M. Jambrecina (2002). Wilderness and cultural landscapes: Shifting management emphases in the Tasmanian Wilderness World Heritage Area. *Australian Geographer*, 33(2), 125–139.

Russell, K. (2001). What is wilderness therapy? *Journal of Experiential Education*, 24, 70–80.

Russell, K. (2005). Two years later: A qualitative assessment of youth well-being and the role of after care in outdoor behavioral healthcare treatment. *Child & Youth Care Forum*, 34, 209–239.

Russell, K. & M. Walsh (2011). An exploratory study of a wilderness adventure program for young offenders. *Journal of Experiential Education*, 33, 398–342.

Rutko, E. & J. Gillespie (2013). Where's the wilderness in wilderness therapy? *Journal of Experiential Education*, 36, 218–235.

Ryan, J. (2000). "Hunting with the camera": Photography, wildlife, and colonialism in Africa. In C. Philo and C. Wilbert (eds), *Animal places, beastly spaces* (pp. 205–223). London: Routledge.

Sæþórsdóttir, A. D., M. Hall, & J. Saarinen (2011). Making wilderness: Tourism and the history of the wilderness idea in Iceland. *Polar Geography*, 34, 249–273.

Salwen, P. (1989). *Galapagos: The lost paradise*. New York: Mallard Press.

Sandilands, C. (1999). Domestic politics, multiculturalism, wilderness, and the desire for Canada. *Space & Culture*, 4, 169–186.

Sandilands, C. (2005). Where the mountain men meet the lesbian rangers: Gender, nation and nature in the Rocky Mountain National Parks. In M. Hessing, R. Raglon, & C. Sandilands (eds), *This elusive land: Women and the Canadian environment* (pp. 142–162). Vancouver: University of British Columbia Press.

Sandlos, J. (2008). *Hunters at the margin: Native people and wildlife conservation in Northwest Territories*. Vancouver: UBC Press.

Sanford, W. (2007). Pinned on Karma rock: Whitewater kayaking as religious experience. *Journal of the American Academy of Religion*, 75, 875–895.

Satterfield, T. (2003). *Anatomy of a conflict: Identity, knowledge, and emotion in old-growth forests*. East Lansing, MI: Michigan State University Press.

Schaffer, K. (1988). *Women and the bush: Forces of desire in the Australian cultural tradition*. New York: Cambridge University Press.

Schama, S. (1995). *Landscape and memory*. London: Harper Collins.

Schmidt, C. & D. Little (2007). Qualitative insights into leisure as a spiritual experience. *Journal of Leisure Research*, 39, 222–247.

Schmidt, C. J. (1979). The guided tour: Insulated adventure. *Urban Life*, 7, 441–467.

Schwartz, K. Z. S. (2005). Wild horses in a "European wilderness": Imagining sustainable development in the post-Communist countryside. *Cultural Geographies*, 12, 292–320.

Scott, K. D. (2003). Popularizing science and nature programming: The role of "spectacle" in contemporary wildlife documentary. *Journal of Popular Film and Television*, 31(1), 29–35.

Sharpe, E. (2005). Delivering communitas: Wilderness adventure and the making of community. *Journal of Leisure Research*, 37, 255–280.

Shetler, J. B. (2006). *Imagining Serengeti: A history of landscape memory in Tanzania from earliest times to the present*. Athens, OH: Ohio University Press.

Silva, J. A. & L. K. Khatiwada (2014). Transforming conservation into cash? Nature tourism in southern Africa. *Africa Today*, 6(1), 17–45.

Simon, G. L. & P. S. Alagona (2013). Contradictions at the confluence of commerce, consumption and conservation; or, an REI shopper camps in the forest, does anyone notice? *Geoforum*, 45, 325–336.

Sletto, B. (2011). Conservation planning, boundary-making and border terrains: The desire for forest and order in the Gran Sabana, Venezuela. *Geoforum*, 42, 197–210.

Smith, K. K. (2008). *What is Africa to me? Wilderness and black thought, 1860–1930*. In M. Nelson & J. B. Callicott (eds), *The wilderness debate rages on: Continuing the great new wilderness debate* (pp. 300–324). Athens, GA: The University of Georgia Press.

Solnit, R. (2001). *Wanderlust: A history of walking*. London: Verso.

Sontag, S. (1979). *On photography*. New York: Penguin.

Soper, K. (1995). *What is nature? Culture, politics, and the non-human*. Oxford: Blackwell.

Soule, M. & R. Noss (1998). Rewilding and biodiversity: Complementary goals for continental conservation. *Wild Earth*, 7(3), 19–28.

Spence, M. D. (1999). *Dispossessing the wilderness: Indian removal and the making of the national parks.* New York: Oxford University Press.

Spenceley, A. (2005). Nature-based tourism and environmental sustainability in South Africa. *Journal of Sustainable Tourism*, 13(2), 136–170.

Spenceley, A. (2008). *Responsible tourism: Critical issues for conservation and development.* London: Earthscan.

Stebbins, R. A. (1992). *Amateurs, professionals, and serious leisure.* Montreal: McGill Queen's University Press.

Stephen, L. C. (2010). "At last the family is together": Reproductive futurism in *March of the Penguins. Social Identities: Journal for the Study of Race, Nation and Culture*, 16(1), 103–118.

Stevens, S. (1997). *Conservation through cultural survival: Indigenous peoples and protected areas.* Washington, DC: Island Press.

Stoddart, M. (2012). *Making meaning out of mountains: The political ecology of skiing.* Vancouver: UBC Press.

Stoll, M. (2007). Religion "irradiates" the wilderness. In M. Lewis (ed.), *American wilderness: A new history* (pp. 35–53). New York: Oxford University Press.

Stoll, S. (2007). Farm against forest. In M. Lewis (ed.), *American wilderness: A new history* (pp. 55–72). New York: Oxford University Press.

Stradling, D. (2004). *Conservation in the progressive era: Classic texts.* Seattle, WA: University of Washington Press.

Stringer, L. A. & L. McAvoy (1992). The need for something different: Spirituality and wilderness adventure. *Journal of Experiential Education*, 15, 13–21.

Suchet, S. (2002). "Totally wild?" Colonising discourses, indigenous knowledges and managing wildlife. *Australian Geographer*, 33, 141–157.

Sundberg, J. (1999). NGO landscapes in the Maya Biosphere Reserve, Guatemala. *The Geographical Review*, 88(3), 388–412.

Sundberg, J. (2003). Conservation and democratization: Constituting citizenship in the Maya Biosphere Reserve, Guatemala. *Political Geography*, 22, 715–740.

Sundberg, J. (2006). Conservation encounters: Transculturation in the "contact zones" of empire. *Cultural Geographies*, 13, 239–265.

Sutter, P. (2009). *Driven wild: How the fight against automobiles launched the modern wilderness movement.* Seattle: University of Washington Press.

Swarbrooke, J., C. Beard, S. Leckie, & G. Pomfrett (2003). *Adventure tourism: The new frontier.* New York: Routledge.

Swatton, A. & T. Potter (1998). The personal growth of outstanding canoeists resulting from extended solo canoe expeditions. *Pathways: The Ontario Journal of Outdoor Education*, 9, 13–16.

Sweatman, M. & P. Heintzman (2004). The perceived impact of outdoor residential camp experience on the spirituality of youth. *World Leisure Journal*, 46, 23–31.

Swyngedouw, E. (1999). Modernity and hybridity: Nature, *regeneracionismo*, and the production of the Spanish waterscape, 1890–1930. *Annals of the Association of American Geographers*, 89(3), 443–465.

Takeda, L. & I. Røpke (2010). Power and contestation in collaborative ecosystem-based management: The case of Haida Gwaii. *Ecological Economics*, 70, 178–188.
Taylor, J. E., A. Yúnez-Naude, G. Dyer, S. Ardila, & M. Stewart (2003). The economics of "eco-tourism": A Galapagos Island economy-wide perspective. *Economic Development and Cultural Change*, 51(4), 977–997.
Taylor, K. & J. Lennon (2011). Cultural landscapes: A bridge between culture and nature? *International Journal of Heritage Studies*, 17, 537–554.
Terborgh, J., C. van Schaik, L. Davenport, & M. Rao (2002). *Making parks work: Strategies for preserving tropical nature*. Washington, DC: Island Press.
"The death of Cecil the Lion" (2015, July 31). *New York Times*. Retrieved from www.nytimes.com/2015/07/31/opinion/the-death-of-cecil-the-lion.html?_r=2 (last accessed December 2, 2015).
Thomas, C. D. (1990). What do real population dynamics tell us about minimum viable population sizes? *Conservation Biology*, 4, 324–327.
Thoreau, H. D. (1962 [1854]). *Walden*. New York: Time Incorporated.
Thoreau, H. D. (1992 [1862]). *Walking*. Bedford, MA: Applewood Books.
Thoreau, H. D. (2007). *Walking*. Rockville, MD: ARC Manor.
Trainor, S. F. & R. B. Norgaard (1999). Recreation fees in the context of wilderness values. *Journal of Park and Recreation Administration*, 17, 100–115.
Tremblay, P. (2001). Wildlife tourism consumption: Consumptive or non-consumptive? *International Journal of Tourism Research*, 3, 81–86.
Turner, J. (1996). *The abstract wild*. Tempe, AZ: University of Arizona Press.
Turner, J. M. (2012). *The promise of wilderness: American environmental politics since 1964*. Seattle, WA: University of Washington Press.
Turner, V. & E. Turner (1978). *Image and pilgrimage in Christian culture*. New York: Columbia University Press.
Urry, J. (1990). *The tourist gaze: Leisure and travel in contemporary societies*. London: SAGE.
US Wilderness Act, Section 2 (1964). *Public Law 88–557 (16 USC 1131–1136)*. 88th Congress, Second Session, 3 September 1964.
Valdivia, G., W. Wolfort, & F. Lu (2014). Border crossings: New geographies of protection and production in the Galapagos Islands. *Annals of the Association of American Geographers*, 104(3), 686–701.
Vander Kloet, M. (2009). A trip to the Co-op: The production, consumption and salvation of Canadian wilderness. *International Journal of Canadian Studies*, 39–40, 231–251.
Van Hoven, B. (2011). Multi-sensory tourism in the Great Bear Rainforest. *Landabrefid*, 25, 31–49.
Van Schyndel, N. (2014). *Becoming wild*. Halfmoon Bay, BC: Caitlin Press.
Van Slyck, A. (2006). *A manufactured wilderness: Summer camps and the shaping of American youth, 1880–1960*. Minneapolis: University of Minnesota Press.
Varley, P. (2011). Sea kayakers at the margins: The liminoid character of contemporary adventures. *Leisure Studies*, 30, 85–98.
Vest, J. H. (1985). Will-of-the-land: Wilderness among primal Indo-Europeans. *Environmental Review*, 9, 323–329.

Vining, J. (2003). The connection to other animals and caring for nature. *Human Ecology Review*, 10, 87–99.

Waitt, G. (1997). Selling paradise and adventure: Representations of landscape in the tourist advertising of Australia. *Australian Geographical Studies*, 35(1), 47–60.

Waitt, G. & L. Cook (2007). Leaving nothing but ripples on the water: Performing ecotourism natures. *Social & Cultural Geography*, 8, 535–550.

Wallace, G. (1993). Wildlands and ecotourism in Latin America: Investing in protected areas. *Journal of Forestry*, 91, 37–40.

Walpole, J. M. (2001). Feeding dragons in Komodo National Park: A tourism tool with conservation complications. *Animal Conservation*, 4, 67–73.

Washabaugh, W. & C. Washabaugh (2000). *Deep trout: Angling in popular culture.* Oxford: Berg.

Watkins, G. & F. Cruz (2007). *Galápagos at risk: A socioeconomic analysis of the situation in the archipelago.* Puerto Ayora, Galápagos, Ecuador: Charles Darwin Foundation.

Watson, A., L. Alessa, & B. Glaspell (2003). The relationship between traditional ecological knowledge, evolving cultures, and wilderness protection in the circumpolar north. *Conservation Ecology*, 8(2). Available at: www.consecol.org.vol8/iss1/art2/ (last accessed January 19, 2016).

Wattchow, B. & M. Brown (2011). *A pedagogy of place: Outdoor education for a changing world.* Monash University Publishing. Available at: http://books.publishing.monash.edu/apps/bookworm/view/A+Pedagogy+of+Place/131/pp10000a.xhtml (last accessed December 2, 2015).

Wellock, T. R. (2007). *Preserving the nation: The conservation and environmental movements 1870–2000.* Wheeling, IL: Harlan Davison.

West, P. & J. G. Carrier (2004). Ecotourism and authenticity: Getting away from it all. *Current Anthropology*, 45(4), 483–491.

West, P., J. Igoe, & D. Brockington (2006). Parks and peoples: The social impact of protected areas. *Annual Review of Anthropology*, 35, 251–277.

Whatmore, S. (1999). Nature culture. In P. Cloke, M. Crang, & M. Goodwin (eds), *Introducing human geographies* (pp. 4–11). London: Arnold.

Whatmore, S. (2006). Materialist returns: Practising cultural geography in and for a more-than-human world. *Cultural Geographies*, 13, 600–609.

Whatmore, S. & L. Thorne (1998). Wild(er)ness: Reconfiguring the geographies of wildlife. *Transactions of the Institute of British Geographers*, 23(4), 435–454.

White, D. D. & J. Hendee (2000). Primal hypotheses: The relationship between naturalness, solitude, and the wilderness experience benefits of development of self, development of community, and spiritual development. In S. F. McCool, D. N. Cole, W. T. Borrie, & J. O'Loughlin (Comps.), *Wilderness science in a time of change conference, Vol. 3. Wilderness as a place for scientific inquiry* (pp. 223–227). USDA Forest Service Proceedings RMRS-P-15. Ogden, UT: US Dept. of Agriculture, Forest Service, Rocky Mountain Research Station.

White, L. & B. King (2009). The Great Barrier Reef Marine Park: Natural wonder and World Heritage Area. In W. Frost & C. M. Hall (eds), *Tourism and national parks: International perspectives on development, histories and change* (pp. 114–127). London, UK: Routledge.

Whittington, A. (2006). Challenging girls' contructions of femininity in the outdoors. *Journal of Experiental Education*, 28(3), 205–221.

Wiley, E. (2002). Wilderness theatre: Environmental tourism and Cajun swamp tours. *The Drama Review*, 46(3), 118–131.

Williams, B. (2000). The treatment of adolescent populations: An institutional vs. a wilderness setting. *Journal of Child and Adolescent Group Therapy*, 10, 47–56.

Williams, K. & D. Harvey (2001). Transcendent experience in forest environments. *Journal of Environmental Psychology*, 21, 249–260.

Williams, R. (1983). *Keywords*. New York: Oxford University Press.

Wilshusen, P. R., S. R. Brechin, C. L. Fortwangler, & P. C. West (2003). Contested nature: Conservation and development at the turn of the twenty-first century. In S. R. Brechin, P. R. Wilshusen, C. L. Fortwangler, and P. C. West (eds), *Contested nature: Promoting international biodiversity conservation with social justice in the twenty-first century* (pp. 1–22). Albany, NY: State University of New York Press.

Wilson, E. O. (1984). *Biophilia*. Cambridge, MA: Harvard University Press.

Wilson, J. & M. Lipsey (2000). Wilderness challenge programs for delinquent youth: A meta-analysis of outcome evaluations. *Evaluation and Program Planning*, 23, 1–12.

Wolmer, W., J. Chaumba, & I. Scoones (2004). Wildlife management and land reform in southeastern Zimbabwe: A compatible pairing or a contradiction in terms? *Geoforum*, 35, 87–98.

Woodward, R. (2000). Warrior heroes and little green men: Soldiers, military training and the construction of rural masculinities. *Rural Sociology*, 65, 640–657.

World Conservation Union (n.d.). Tourism and the environment. Available at: https://portals.iucn.org/library/efiles/html/tourism/section5.html (last accessed December 2, 2015).

Wuerthner, G., E. Crist, & T. Butler (2014). *Keeping the wild: Against the domestication of earth*. Washington, DC: Island Press.

Wylie, J. (2005). A single day's walking: Narrating self and landscape on the South West Coast Path. *Transactions of the Institute of British Geographers*, 30, 234–247.

Young, Z., G. Makoni, & S. Boehmer-Christiansen (2001). Green aid in India and Zimbabwe—conserving whose community? *Geoforum*, 32, 299–318.

Zanotti, L. & J. Chernela (2008). Conflicting cultures of nature: Ecotourism, education and the Kayapó of the Brazilian Amazon. *Tourism Geographies*, 10(4), 495–521.

Zimmer, O. (1998). In search of national identity: Alpine landscape and the reconstruction of the Swiss nation. *Comparative Studies in Society and History*, 40(4), 637–664.

Zimmerer, K. (2006). Cultural ecology: At the interface with political ecology–the new geographies of environmental conservation and globalization. *Progress in Human Geography*, 30(1), 63–78.

Zimmerer, K. S. & K. R. Young (1998). *Nature's geography: New lessons for conservation in developing countries*. Madison: University of Wisconsin Press.

Zmelik, K., S. Schindler, & T. Wrbka (2011). The European green belt: International collaboration in biodiversity research and nature conservation along the former Iron Curtain. *The European Journal of Social Science Research*, 24(3), 273–294.

Index

Bold type represents an entry in a table or box.